AIRLINE
ROUTE PLANNING

AIRLINE
ROUTE PLANNING

John H. H. Grover
FRAeS, MRIN, MBAC (Hon. Life)

BSP PROFESSIONAL BOOKS

OXFORD LONDON EDINBURGH

BOSTON MELBOURNE

First published 1990

British Library
Cataloguing in Publication Data
Grover, John H.H.
  Airline route planning.
  1. Air services.   Routes.   Planning.
  I. Title
  387.7′4042

ISBN 0-632-02324-4

BSP Professional Books
A division of Blackwell Scientific
  Publications Ltd
Editorial Offices:
Osney Mead, Oxford OX2 0EL
  (Orders: Tel. 0865 240201)
8 John Street, London WC1N 2ES
23 Ainslie Place, Edinburgh EH3 6AJ
3 Cambridge Center, Suite 208, Cambridge
  MA 02142, USA
107 Barry Street, Carlton, Victoria 3053,
  Australia

Set by DP Photosetting, Aylesbury, Bucks
Printed in Great Britain by
St Edmundsbury Press Ltd
Bury St Edmunds, Suffolk

Acknowledgements

The UK Air Navigation Order definitions given in Appendix B are reproduced from the Air Navigation Order: SI 1985/1643 with permission of the Controller of Her Majesty's Stationery Office.

The ICAO definitions given in Appendix B are reproduced from ICAO Publications Doc 8168-OPS/611, Part 1 with permission of ICAO.

Contents

Foreword

by
Air Marshal Sir Ian Pedder KCB, OBE, DFC
Chairman, Dan-Air Services Limited

My introduction to route planning took place early in 1945 when I was instructed to fly cross-country in a Tiger Moth from Scone near Perth to Dunkeld, Kirremuir and return to Scone. It must have been one of those beautiful Scottish days with unlimited visibility because my log-book shows that I succeeded in arriving back at Scone fifty minutes after takeoff.

Of course, the route planning element of *ab initio* pilot navigation training at that time was prepared by the instructional staff. Primarily, its aim was to ensure that the students could get round the cross-country without running out of fuel or running into high ground and that they were presented with plenty of railway viaducts, major crossroads and lozenge-shaped lochs en route. These features had to be correctly identified by Mark I eyeballs and correlated with the map and flight plan by the personalised computer carried between the ears. It was all pretty simple; if the weather was bad you did not go, air traffic slots had not been invented so you just queued for takeoff and jockeyed for position on the downwind leg and, if it seemed as if you might be running out of fuel before you got to your destination, adequate alternative airfields were thick on the ground.

So much for nostalgia! As we enter the 1990s, we should pause to recall the extraordinary standards which have been achieved in the field of aircraft operations. Despite the dramatic and regrettable accidents which occur from time-to-time, the world's airlines continue to offer the surest way of moving travellers between countries and across continents. With consistent safe operation as the dominant criterion, the management of today's complex fleet of high-performance airliners in a congested and environmentally-sensitive world, is probably one of the most challenging of modern enterprises. A vital element of this enterprise is route planning.

The route planning function in a major airline operating on international routes will probably be conducted by a team headed by a senior operations manager. The team will need a thorough understanding of the commercial objectives before addressing the general characteristics of the equipment, airports and air traffic management system to be taken into account. As with all problem-solving, the quality of the solution will depend on the expertise of the team and the precision and currency of the factors which contribute to their calculations.

The planning techniques which began to evolve in the 1930s, gathered pace with the impetus of the worldwide application of military air power in the 1940s and post-war civil aviation have arrived at a stage of maturity and sophistication where it is timely to produce an up-to-date book covering all the elements involved.

This is a challenge for any author, made all the more demanding if he elects to aim his material at the general reader of aviation affairs as well as the specialist.

Indeed, this book illuminates the many facets of airline route planning in a thoughtful and practical way. Whilst it is written in a European context, its application is worldwide. It may be used as a general reference book by experts or for detailed guidance by newcomers to the subject.

I believe that John Grover has met the challenge admirably and he and his publishers, Blackwell Scientific Publications, are to be congratulated at the successful outcome of their enterprise.

Dan-Air Services Limited
London

November 1989

Preface

In using the title 'Airline Route Planning' for this book I am deliberately taking this in its broadest sense. That is to say, what is involved overall in the planning and operation of airline routes from the provision of flight documentation for flight crew use, the calculation of commercial operating estimates, the evaluation of new aircraft types, through to the monitoring of operating techniques and flight procedures.

This book is written for all levels and aspects of airline management, from the Chairman down to the youngest management trainee being groomed for stardom. It seeks to show non-flight staff the problems that can concern flight deck staff and the rules that must be obeyed. It also is intended to show flight deck staff some of the considerations that are behind their flight documentation. It therefore tries to bring about a greater understanding between the various departments within an airline, hopefully resulting in a concomitant increase in overall efficiency. It may even be a useful tool for those considering the setting up of a new airline.

As in my previous, and companion, book (*Handbook of Aircraft Performance*) I have used as my example aircraft the BAe 146-100. This not only is, I believe, the first transport aircraft to be certificated to the European Joint Airworthiness Requirements (JAR 25) but it is also certificated to the US FAR 25, upon which JAR 25 is based. It is, therefore, a representative example of current thinking and practice. JAR 25 are currently phasing out British Civil Airworthiness Requirements and all new designs will be subject to the former requirements.

The majority of the western world's airliners in the passenger transport category are certificated to FAR 25. It is logical, therefore, to continue this book from the companion volume by using as an example a type that represents this majority as well as future European aircraft. To understand FARs and JARs for one type is to make it simple enough to understand another in terms of rules of operation.

I have, I hope, shown why I have used this modern design. Another great advantage is the fact that British Aerospace, Hatfield, are close at hand – unlike Boeing, Seattle. This makes for ease of reference, in case of any query.

An important point that I would like to make, and which will be repeated in the text, concerns the use of the illustrations and data contained in this book. Under *no* circumstances should any such data be used operationally. It is there for illustration purposes only, and does not purport to be accurate, although I have tried to prepare such data accurately. I do believe that the actual principles – which are what matters in this context – are accurate, though. And it should not be forgotten that practically everything in aviation is liable to change; what was accurate this morning may no longer be so this afternoon! Therefore I would like to draw the reader's attention to the fact that the contents of this book are as were applicable on 10 March 1989.

Finally, I must give credit to where it is due. Firstly I would like to thank Air Marshal Sir Ian Pedder, Chairman of Dan-Air Services Ltd, for providing the Foreword to this book. I am greatly indebted to Mr Malcolm Galbraith, and Mr Ron Bailey at British Aerospace, Hatfield, for their ready assistance in the provision of data, and other information. I would also like to thank Captain L.J. Buist, of Dan-Air Services Ltd for his help in giving me an insight into the latest technology as it affects route planning, and for providing me with many illustrations. I must also thank Mr Ian Strike, of one of the UK's 'youngest' airlines, Paramount Airways Ltd, for his help and advice, not to say time. And finally, to British Airways AERAD, for their provision of much up-to-date illustrative material – for which, my grateful thanks.

Bradford on Avon   *September 1989*

1:   What is Route Planning?

Let us start off by being clear about what we are considering and discussing. There is a potential area of possible confusion centred around the procedure known as 'flight planning'. In fact, flight planning is closely allied to 'route planning', and may even be considered to be the offspring of the latter. So, let us define flight planning first.

Flight Planning
Flight planning is a procedure that must be carried out before each and every commercial flight – or at least should be. It involves the calculation of the fuel required for the proposed flight, this being the fuel expected to be burned (the 'burnoff') from the start-up at the departure airport to the landing at the destination, including all taxying (although if the landing weight is critical at the destination it is necessary to include in the landing weight the weight of the fuel that will be burned during the taxying-in). A specified percentage of burnoff must be added to the flight burnoff (takeoff to landing) – normally around 5% – for contingencies, navigational errors, route wind changes from those forecast, etc. In addition, the fuel burnoff must be calculated for a possible diversion from the destination to a nominated 'alternate', plus the contingency percentage on *this* possible burnoff, as above. Then a fixed allowance must be added to cover a period of 'holding', or 'stacking', while awaiting one's turn to approach and land. The total fuel required is therefore

(Airport 'X' to Airport 'Y') + 5% + (Airport 'Y' to Airport 'Z') (alternate) + 5% + holding + taxy allowance

This total is the required 'sector fuel'.

The sector fuel weight (less the fuel required for taxying-out) must be entered on the 'load sheet' which, together with the aircraft's APS (aircraft prepared for service) weight and the weight of the 'payload', gives the required takeoff weight (Req. TOW). Given this final weight and details of the runway, airport elevation, surface wind and temperature, the significant takeoff speeds – V_1, V_r, V_2, etc. – may be calculated and noted, as required by regulation. Also to be calculated is the time, both for the sector and also between designated points along the route.

The Flight Plan
The 'route time' will vary according to the forecast conditions for the

flight. The total time must be entered on an internationally accepted form known as the flight plan (although sometimes the 'navigation log' can also bear the same title, which can be confusing). Among the various things that must also be entered on the flight plan are the total endurance for the fuel loaded, the total time from 'X' to 'Y', the nominated alternate airport, the estimated, or calculated, time from takeoff to the first required reporting point (RP) on the route, and the times between each RP for the route being flown. All this information, taken from the flight plan, is required for air traffic control purposes. Similar information is also noted on the aircraft's navigation log, on which provision is usually made for expected, and actual times both between the RPs and for the whole distance. Likewise, expected fuel burnoff is similarly recorded, together with actual burnoff and from this, fuel remaining is assessed. Throughout the flight actual fuel remaining is compared with the fuel that was forecast to be remaining, at regular intervals, and thus a 'fuel log' is kept. If the fuel being burned exceeds that forecast for the various RPs along the route it is necessary to continuously recalculate the ETA (estimated time of arrival) fuel, and if this latter falls significantly short of that required an en route landing for refuelling may be indicated. Fortunately, this is a rare contingency, thanks to the efforts of the route planning people; route planning exists to provide, among other things, data that enables the flight planning process to be proceeded expeditiously and accurately.

Route Planning
Route planning can have other titles, and the way it is organised can vary widely. In its simplest form, appropriate to a small airline, it can be a 'man and boy' (or the equivalent in the feminine gender, of course) operation whose function is more or less limited to providing flight information, such as the Route Book volume of the Operations Manual and prepared navigation logs, and the provision of flight time, fuel burnoffs, and payload estimates for a particular route to enable the airline's commercial department to cost, and tender for, a charter. In the prepared navigation logs (or 'plogs') will be found navigational details for the route in question, together with estimated standard times and fuel burnoffs between RPs, or 'waypoints'. These standard fuels, added to the standard diversion fuel estimate and the holding fuel etc. become the minimum sector fuel (MSF) in total. This is the *minimum* fuel that may be in the aircraft's tanks at takeoff; even if the forecast conditions indicate that less is required the fuel load may not be less than the MSF. All of these activities will be discussed in detail later in the book.

Having shown what is, in fact, just about the minimum activity acceptable for the title 'route planning', let us look at the other end of the spectrum, as appropriate to a large airline. The same basic requirements, as outlined above, are used, but by a much larger department or unit. Sometimes, for example, the activities described above will be provided by a unit having considerably more than the 'man and boy' staffing level, and being known as, say, the route facilities department. There could also be the 'flight development' (or flight technical) department whose function is to devise the optimum means of utilising the company's fleet, bearing in mind engine handling procedures, flight instrumentation and flight management installations, flight profiles, and so forth. It will also keep abreast of new aircraft designs and, if required, analyse these for use on the company's existing or proposed route structure (see Chapter 12). If necessary, flight checks will then be made, and recommendations issued to the airline's management. If the new aircraft are then acquired it will be flight development's task to see these into service, in co-operation with the route planning (or route facilities) unit. The latter will then issue the appropriate flight documentation, such as route books, 'plogs', and flight data.

Working with flight development may well be an engineering unit, whose task it will be to devise a maintenance programme and to assess engineering costs and practices for each aircraft type operated by the company. Yet another company unit will establish handling costs and the contract price for fuel, together with landing fees, agency costs, and so on. Handling in this context refers to airport ground assistance and agency activities, usually away from base. Crewing requirements will also need to be taken into account and the end result should produce both an aircraft net hourly operating cost and also a route operating cost.

Let us now illustrate a typical case to show the broad outlines of what can be deemed to be route planning. Such a structure will be assumed throughout this book.

Route Planning Practice

Initially there will be a 'panic', normally emanating from the airline's commercial department. Perhaps this may concern a single charter, or an additional route to be added to the airline's network, or even a multistage service. With the exception of the charter case it can be assumed that the commercial department will have carried out earlier some form of market research which has made them feel that the proposed addition has potential. Somewhat cynically, some may say, the author feels that all too often the commercial people keep their cards too close to their chests, and only call for a technical feasibility study at an unnecessarily late stage. Hence the initial 'panic', because the answers from the route planners often seem to be required at least the day before they are requested! The commercial department also seem to need the maximum possible payload availability, irrespective of whether or not this is likely to be required.

Route Licences

The most likely pattern to emerge is as follows, using a proposed new scheduled service as an example. Hopefully, with the enquiry will come an indication as to which aircraft type in the company's fleet is most likely to be involved. Armed with this information the basic route planners can work out optimum routings for the route leg or legs, together with a selection of suitable alternates. The next stage is to obtain the route statistical weather conditions, namely winds and temperature. From this data minimum fuel loads (MSF) per stage, and thence payloads can be assessed, together with average stage times. If the commercial people like the results at this stage they will then undertake a costing exercise and thereby try to establish a potential cost:profit ratio, using their market research to estimate revenue. If the results still appear to be attractive the next stage in the evolution of the proposed route is the seeking of official approval for the route to be operated, and, where appropriate, for a route licence to be issued. *Now the real fun starts!*

Once an application has been made for a route licence (in those countries where it is required, such as the UK) a number of things happen. The first of these usually is that the application is published in such a way that it comes to the notice of other 'interested parties'. Normally these are other airlines in the country within which the applicant is domiciled. The publication of the application is to allow such parties to object to the application, giving their reasons. For example, they may complain that if the application is granted it will divert some traffic from their own route or routes. Or they may say that they were just about to apply for such a licence themselves. If the applicant is a private airline it is highly likely that the state airline may object on principle. (For instance, when British Airways was a state corporation, it appeared [to the author] to object automatically to nearly every application.)

There is a time limit for objections to be filed, and all objectors must be provided with a copy of the applicant's commercial estimates that must accompany the application (UK). These estimates need to be accurate and detailed; in particular the costings must be well researched. For example, fuel is a very substantial element in the costs of a route. But there is no standard fuel cost, even from the same 'pump'. In the aviation world the cost per US gallon is negotiated and is quoted on the actual, or estimated, annual uplift by the applicant. The objectors will seek to pick holes in the costings, and will use their own route planners to check these with a very fine toothcomb. They will also question the forecast revenue; even the profit from duty free sales on board is included, and may be questioned. Most of this argument normally takes place at a public hearing, which is conducted in many ways like a court of law, although much less formally. (The applicant may be likened to the defence and the objectors to the prosecution.) The proposals, and counter-proposals, equate to evidence; if the hearing board believes the applicant he will be granted the licence (subject to financial stability, resources, and the acceptance of his operational proposals). But if the objectors succeed in discrediting the applicant's figures then the hearing will probably refuse the application. Even so, all is not lost as there is provision for an appeal procedure. Then, if the application is granted, there follows a period of intense operational activity across the whole route planning spectrum.

Already the type of aircraft to be used on the new service(s) will have been determined, prior to the application and hearing. These may be in service with the company at the time, or will have been selected as being the most suitable new aircraft to be acquired. In either case, the fullest route operating procedures and techniques will have to be established and then submitted to the appropriate regulatory enforcement agency – e.g. in the UK, the Flight Operations Inspectorate of the CAA, and in the USA the comparable branch in the FAA. Typical data required will be the proposed 'aerodrome operating minima', airfield takeoff and landing data for each runway involved, routings, and MSF values. This data will normally comprise extra pages for inclusion in the route book, and plogs. (Incidentally, plogs are being increasingly replaced by computer-produced logs.) If a new type of aircraft is involved a new route book will be necessary for the type, *in toto*, and the appropriate crew training will have to be carried out so that the crews involved are properly type-rated and checked. It may also be necessary to fly proving flights, especially if 'difficult' airports are to be involved – e.g. Innsbruck, Austria, and Hong Kong. Only when the regulatory authorities declare themselves to be satisfied may the new service be operated commercially.

On-line Procedures

Another function of route planning can be the monitoring of flight procedures 'on line' – that is to say, in service. For example, it may initially have been decided to operate the new service at a certain engine setting, for economic reasons. Evidence may possible accrue that shows this setting to be less advantageous than another. This evidence must be checked, and if upheld a new route operating procedure be notified by means of a route book amendment and a flying staff instruction (FSI). To illustrate this point, it was once the experience of the author to investigate the validity of operating a certain type of turboprop aircraft at a turbine gas temperature (TGT) of 705°C, allied with a compressor and turbine speed of 14200 rpm. The alternative settings were 730°TGT/14000 rpm. Apparently the engine overhaul cost was lower at the former setting, although the aircraft speed was also lower. A route check was made over several sectors, with a regular change of engine settings from one to the other every 20 minutes. After allowing the engine to stabilise at each change of setting the IAS (indicated airspeed), outside air temperature, fuel flows, altitude, etc. were logged. On return to base these logged values

Table 1.1 Route planning functions.

(a) The provision of flight documentation in simplified, practical form for flight crew use.
(b) The provision of payload, fuel burnoff, and block time estimates for commercial use.
(c) The provision of operational data and supporting evidence in the case of route licence applications.
(d) The monitoring of route flight techniques.
(e) The provision of annual total route fuel uplift requirements.
(f) The evaluation of new aircraft on their merits for possible acquisition by the company, together with new navigational and flight instrumentation systems.
(g) The checking of en route facilities (navigational) for possible changes, modification, withdrawal and so on. The checking of en route airport and ground facilities to ensure compatibility with the aircraft in use, or proposed for use, on each route (landing aids, fuel availability to required specification, ground handling facilities, i.e. for handling passengers, freight etc. and the provision of ground equipment such as stairs, GPU, and so on.)
(h) Where necessary the design and promulgation of suitable instrument approach procedures, where none exist (e.g. in the case of a new airport that is brought into service before any IAPs are published. Genoa 1963, is an example that may be cited; the author prepared the first IAP and flight-checked this, prior to the commencement of a new scheduled service).
(i) The preparation and conduct of any proving flights.
(j) To liaise with the company's other interested departments on all aspects of flight operations, and to act as an adviser to management on such matters.
(k) Diplomatic requirements and clearances.

were plotted for each engine setting, as appropriate, and from each plot it was possible to obtain sector averages of IAS and burnoff for both engine settings. It emerged that, using the 705°TGT/14 200 rpm power setting a sector that took, say, four hours would have only taken 3 hours 40 minutes had the alternative power setting been used. When this saving in time of 20 minutes was communicated to the engineering department it was found that the saving in costs per flight hour was greater than those resulting from the use of the lower TGT and higher rpm combination. From then on the particular aircraft fleet involved was operated at the alternative power setting.

Route Planning's Functions

So, route planning can be a wide-ranging subject, although it may be carried out by a *structure* that varies from airline to airline. It may, therefore, be useful if the scope of route planning itself is tabulated (Table 1.1), and it is in this form that it will be dealt with throughout this book.

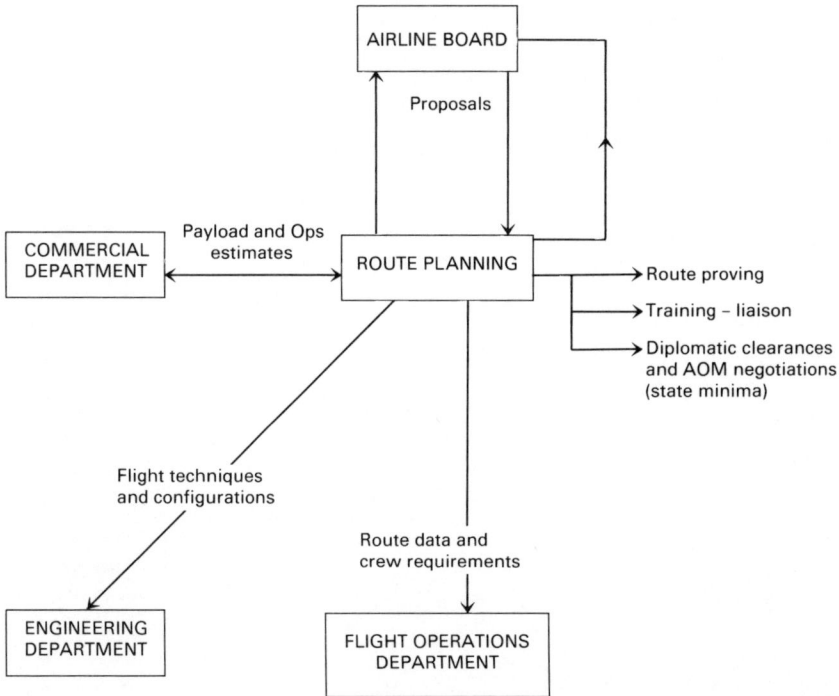

Fig. 1.1 A flow diagram illustrating the route planning function: it could also be a 'large airlines' route planning department lines of communication and responsibility layout.

Never mind about names and departmental titles – this is what the author sees as being the scope of route planning in airline use (and please note the use of the word 'airline', as opposed to 'airways'! An airway is a controlled, corridor-shaped, airspace, along which airline – i.e. air transport undertakings – aircraft fly. This misuse of the word airways is mainly prevalent in the UK and its adherents).

Table 1.1 may not be all-embracing, although it is believed that it is as close to this as is reasonable to assume. Route planning, *per se*, must of necessity be a committee function between several different disciplines and types of expertise. It is what is covered by this 'committee' that is true route planning in its fullest sense. See Fig. 1.1.

2: Route Planning – A Basic Function

In this chaper we will examine one of the most important, and certainly time consuming, functions of an airline's route planning section. In fact, this function probably accounts for a good 50% of the RP section's man hours, especially in the case of a small airline. This is the provision of payload, block time, and block fuel estimates in response to enquiries from the airline's commercial department. Reference has already been made to this in Chapter 1. We will now take a close look at what is involved when an enquiry is received.

Charter Tendering
A phone rings, or a door bursts open, and the game starts! It is the commercial department, and they have just received an invitation to quote for a charter flight from airport 'A' to airport 'B'. They *must* have the answer within 30 minutes, and need three estimates, i.e. for payload, block time and block fuel. The immediate question from route planning must be, of course, what sort of load is involved – i.e. passengers or freight – and some guide as to numbers or weight. This is particularly important in the case of an airline having more than one type of aircraft in its fleet. To be fair, the commercial people normally have a fairly good idea as to what aircraft type will be needed. At the same time, it pays to be certain about this from the very onset of the matter. It is no use providing a study using, say, a B747, only to find that the prospective customer really only needs a B737. Yet some commercial staff will insist on aiming to offer the greatest payload available, irrespective of the customer's actual needs. Mercifully, they normally learn to find out these needs after a few abortive exercises and soon specify the load requirements, at least, to route planning. Often, too, even the aircraft is specified.

Armed with commercial details route planning can start work on preparing a detailed estimate. Not only are particulars of the route required but also the time of the year and of the day. The reason for these two items of information is to take account of the statistical temperatures (for the takeoff and the en route element) that will probably apply to the flight, and also the statistical route winds.

The foregoing meteorological data is normally available either from the national Meteorological Office, in the form of published statistics, or from some other source that publishes such data – e.g. Aerad. Most meteorological offices in other countries publish such data, or can supply it, even for a 'one-off' flight. Usually these statistics are produced in association with a 'standard deviation', which enables the statistical data to be adjusted so as to give an acceptable element of certainty to the value chosen. For example, it is usual to take what is known as the 85% temperatures and route winds when estimating fuel requirements and the 50% values for time and scheduling. (An 85% wind or temperature is one that will only be more adverse on 15% of occasions, while the 50% case is average.) The reason for being so conservative about fuel load is that it would not be either operationally or commercially acceptable to contract for a flight that has any significant element of chance that might result in the required payload not being available on the day of operation, due to higher than expected values of temperature and route headwinds. An 85% certainty of operational values is perfectly acceptable. In the case of time, this can only affect the cost and therefore the profit element. The route average conditions (50%) are acceptable for this, and some contingency is usually allowed for anyway. But, if on the day the actual meteorological values correspond to the, say, 75% case the flight may be operated, although it will take a bit longer and use more fuel. But, had the *whole* estimate been based on the 50% case, the fuel *required* would have exceeded that taken into account for the payload calculation, and very possibly either the payload would have to be reduced on the day, or an en route technical stop would be required for refuelling. The latter would involve substantially increased costs; the former might have breached a contract.

Aircraft Configurations
One further item of information that must be obtained, if possible, from the commercial department concerns the configuration of the aircraft type to be used. Again, the customer may have specified the number of seats that are required, or the type of freight to be carried, or both. It is normal for an airline to be able to operate the various aircraft in its fleet in a variety of configurations – i.e. internal layouts. These range from, say, an all-freight layout having no seats for passengers, to a mixed passenger/cargo configuration, to an all-passenger layout. The latter can also vary, from a one class, high density layout, to a two class arrangement, to a multiclass (e.g. first, 'club' and economy). The total number of seats will vary according to the configuration, and it is most important that the calculations are based on the optimum configuration. For example, it is not in the airline's best interests if, say, the estimates are

prepared for a high density configuration when the customer, in fact, really only requires perhaps 75% of the seats available. It would mean that, if the flight, or series of flights, operated thus, the passengers would be travelling in less comfort than they could have had, if the configuration offered and contracted had 25% less seats, but was distributed over the same cabin area. It is not a function of route planning to question such matters, but simply to assume that the commercial department know what is required by the customer and thus place their enquiry accordingly. Normally this will be part of the commercial 'learning curve', and their enquiry will be specific, naming a certain configuration. A possible exception could be when a marginal payload availability arises, either through airfield-limited takeoff or landing weights or sector fuel load requirements or both. In either of these cases there may be a payload loss in the stated configuration that can be improved upon by the use of a layout with fewer seats and therefore a lighter aircraft. It is all a matter of finding out what the customer *really* needs – early!

Aircraft Prepared for Service (APS) Weights

Route planning will hold copies of what are referred to as 'aircraft

Table 2.1 The APS weight statement. (*Note* for illustration purposes only.)

Aircraft type 'X' Registration G-ZXYZ Configuration: high density, 150 pax	
Empty weight	52 000 kg
Unusable fuel (in pipes, etc.)	200 kg
Oil	500 kg
Seats	
30 × doubles at 20 kg	600 kg
30 × triples at 25 kg	750 kg
Pax service items (e.g. pillows, etc.)	150 kg
Catering (standard)*	170 kg
Bar boxes and contents	145 kg
Fluids (e.g. toilets etc.)	200 kg
Crew (2M, 3F)	345 kg
Navigation equipment (manuals, charts)	15 kg
Crew baggage	50 kg
APS weight	55 125 kg

* This may have to be varied to suit a specific request. In such cases the APS weight will be adjusted accordingly.

prepared for service' forms (APS forms) for each type of aircraft, and each individual aircraft, operated by the airline. These APS forms give the weight for each aircraft in each configuration used, and are normally the total weight of each aircraft excluding usable fuel and payload only. There may be slight variations from one airline to another, as a result of policy, but in general the APS weight statement will look similar to Table 2.1.

Payload
We now have the weight of the aircraft awaiting only the addition of the fuel required and the payload. This latter is limited structurally by the aircraft's maximum zero fuel weight (MZFW), i.e. the total weight of the aircraft (APS) plus the payload, but *excluding* the required sector fuel. Taking the APS weight quoted above, viz: 55 125 kg, and assuming that the certificated MZFW is 69 000 kg, the maximum permissible payload is 69 000 − 55 125 = 13 875 kg. This is the highest payload that may be carried, and can only be increased by a lightening exercise on the APS weight − e.g. reduce the weight of catering and bar. Whatever can be pared off the APS weight may be added to the MZFW limited payload. We must now see what payload can, in fact, be offered.

It is often usual to work back from the maximum landing weight, using an approximation for the required fuel burnoff, in order to establish the average weight of the aircraft for the flight. The idea is to find the optimum altitude for the sector to be flown, and thus to calculate the climb, cruise, and descent fuel burnoff, (normally from engines on to engines off), plus the laid down reserve fuel, in order to establish accurately the required fuel load. Also obtained from the calculation is the stage time.

Proceeding further along the path towards an answer for the commercial people we have now obtained a sector burnoff and time. To these figures must be added a 'block' allowance. For fuel, this comprises sufficient to taxy out from the ramp to the runway, and then to taxy in after landing. For time, it is normal to allow a fixed taxy out time (say 10 minutes) and taxy in time (say 5 minutes). The total time is added to the stage time to give block time. Using a known taxy fuel flow rate, the taxy out fuel and the taxy in allowance can be estimated for each time.

Note
In the case of modern aircraft a maximum ramp weight is also laid down.

Maximum ramp weight − MTWA = maximum taxy out fuel available without weight considerations

We now have a block time and a block fuel, and these form the basis for the cost calculation, upon which the quotation is made to the customer. But we still need a payload to offer as well. To obtain this we must calculate the regulated takeoff weight (RTOW) as permitted by the departure airport's runway and altitude above sea level, and also the regulated landing weight (RLW), both according to the expected met. conditions.

First the RTOW is calculated, and then the RLW. If neither limit – i.e. if the aircraft may takeoff at max. permissible takeoff weight (max. takeoff weight authorised or MTWA) and may land at max. permissible landing weight (max. landing weight authorised or MLWA), then it is *probable* that the MZFW payload may be offered. This may be checked by adding the total fuel load (min. sector fuel or MSF) less the taxy-out allowance, to the MZFW; the total weight thus found may not exceed the MTWA. If it does then the payload offered must be reduced by the amount of the excess weight so as to bring the total of the ZFW + MSF down to the MTWA. Likewise, assuming that the takeoff is at MTWA, then the takeoff weight less the burnoff for the stage must not exceed the MLWA. If the all-up weight at landing exceeds the MLWA then the payload must be further reduced so as to ensure that the landing weight does not exceed the MLWA. In short, the payload available must always be such so as to ensure that any takeoff or landing weight limits, are not exceeded and this also applies to those cases where the takeoff weight is less than the MTWA due to runway or climb limits, and where the landing weight is limited for similar considerations. The MSF may not, *under any circumstances*, be reduced and therefore the payload must be adjusted so that the MTWA (or RTOW) and MLWA/RLW values are not exceeded. (See Figs 2.1 and 2.2.)

Assume that the MTWA of the aircraft is 80 000 kg, and that the takeoff RTOW is limited to 75 000 kg. The required MSF is 12 000 kg, so that the *required* TOW is 69 000 kg (MZFW) plus 12 000 kg fuel which equals 81 000 kg. But the RTOW is limited to 75 000 kg, so that the required payload (13 875 kg) must be reduced by 6000 kg. 13 875–6000 kg = 7875 kg and this is a substantial reduction when looking at the MZFW limited payload. In fact, at an average passenger unit weight (say 85 kg which includes required baggage allowance), 7875 kg = only 93 pax! But, if there had been no runway or other takeoff limit, we could take the MZFW 69 000 kg, add the MSF 12 000 kg, and find the *required* TOW of 81 000 kg. But MTWA is 80 000 kg, so that the payload must be reduced by 1000 kg to achieve an acceptable TOW of 80 000 kg. The payload will therefore be 12 875 kg, and this equals 151 pax units, so that the full load of 150 seats can be offered. The only possible drawback is in the hypothetical case

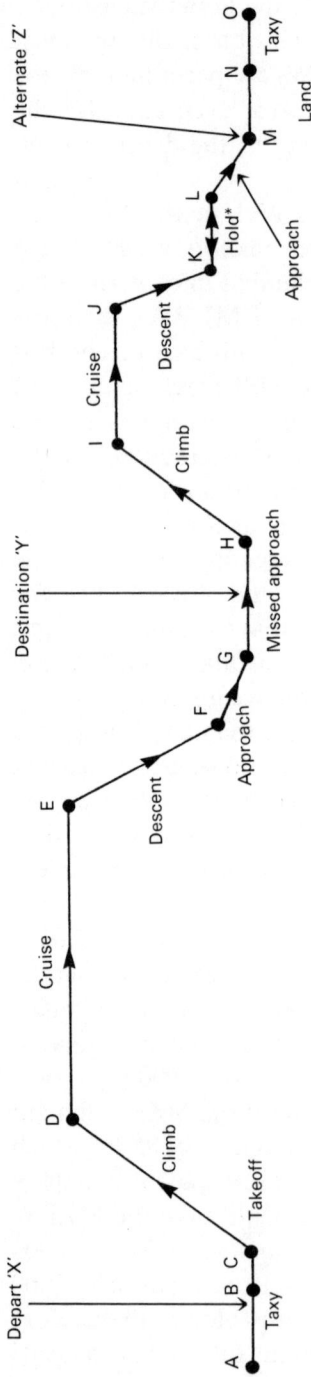

Fuel required = Total A to O *plus* 5% (B-K) + (L–N)

* Holding: 45 minutes at (normally) 8000 ft for turboprops
  30 minutes at (normally) 1500 ft for turbojets
  Both in ISA

Burnoff = Total fuel burned from A to G, plus Landing and Taxy-in at 'Y' (for costing). (A–B) may be
  omitted for takeoff weight calculations.

Fig. 2.1   Schematic presentation of sector fuel requirements. The total fuel at A is the greater of either minimum sector fuel as laid down by the
operator or that required under the forecast meteorological conditions prior to takeoff.

Fig. 2.2 Predeparture after flight planning. A BAe 146–100 of Dan-Air at London (Gatwick) airport. (Courtesy of Dan-Air Services Ltd.)

where all the passengers are males, which would give a standard pax unit weight of 90 kg, and thus the available payload weight 12 875 would yield only 143 pax.

Note
All runway or airfield limited takeoff weights are based on the aircraft weight at brake release for the takeoff. The aircraft weight at start-up may exceed the RTOW by the amount of fuel required to taxi to the brake release point. In most modern aircraft a limiting ramp weight – i.e. the MTWA plus a *maximum* taxy-out allowance – is scheduled.

Some Examples
It may be of value if we illustrate a few payload calculations based on the aircraft's leading data and the weight of the burnoff between 'A' and 'B' (from brake release to landing). We already have the following: MTWA 80 000 kg, MLWA 78 000 kg, MZFW 69 000 kg, and APS weight 55 125 kg. Let us assume that the burnoff from brake release to landing is 7000 kg, and that the MSF is 12 000 kg. From these weights we can calculate a number of varying payload weights, according to varying conditions. MTWA or RTOW values are the brake release weights, and RLW is the landing weight limited by the destination airfield 'B'.

Example 1

	RTOW	75 000 kg	
less	MSF	12 000 kg	(at brake release point)
	ZFW	63 000 kg	
less	APS	55 125 kg	
	P/L	7 875 kg	

Limiting factors: RTOW + MSF (can MFS be reduced by using a closer alternate?)

Example 2

	RTOW	80 000 kg	
less	MSF	12 000 kg	(at brake release point)
	ZFW	68 000 kg	
less	APS	55 125 kg	
	P/L	12 875 kg	

Limiting factor: MSF (seek closer alternate thus reducing MSF?)

Example 3

	RLW	76 900 kg	
plus	Burnoff	7 000 kg	(from brake release point)
Required	TOW	83 900 kg	(but MTWA 80 000 kg)
Revised	RTOW	80 000 kg	
less	MSF	12 000 kg	
	ZFW	68 000 kg	
less	APS	55 125 kg	
	P/L	12 875 kg	

Limiting factors: MTWA + MSF (see Example 2 or reduce APS?)

Example 4

	RLW	78 000 kg	(MLWA)
plus	Burnoff	7 000 kg	(from brake release point)
Required	TOW	85 000 kg	(but MTWA 80 000 kg)
	MTWA	80 000 kg	
less	MSF	12 000 kg	
	ZFW	68 000 kg	
less	APS	55 125 kg	
	P/L	12 875 kg	

Limiting factors: MLWA, MTWA, and burnoff (see Example 3)

Example 5

	RTOW	80 000 kg	
less	MSF	11 000 kg	(at brake release point)
	ZFW	69 000 kg	(MZFW)
less	APS	55 125 kg	
	P/L	13 875 kg	

Limiting factor: MZFW (reduce APS?)

Example 6

	RLW	72 000 kg
plus	Burnoff	7 000 kg
Required TOW		79 000 kg
less	MSF	12 000 kg
	ZFW	67 000 kg
less	APS	55 125 kg
	P/L	11 875 kg

Limiting factors: MLWA and MSF (choose closer alternate, or reduce APS?)

In all the above examples the actual weights used do not refer to any particular aircraft and they are quoted solely to illustrate the various examples. Where potential means of improving the payloads are quoted in brackets after the limiting factor(s), these are simply suggestions. Normally a declared alternate is one that results in the carriage of minimum reserve fuel anyway, as fuel costs money to carry simply as a function of its weight. However, in certain cases it may be an airline's policy to opt for a more distant alternate due to commercial or administrative reasons, e.g. it may already be a part of the airline's route network, and therefore have its own 'in-house' facilities.

It has been known in the past for an airline to resort to what, in the author's opinion, amounts to an undesirable practice when there is a landing weight limit at the intended destination, due to landing distance available. The idea is to reverse the normal, accepted procedure, and declare the alternate as the intended destination, and vice versa. This practice then results in the landing distance available at the *intended* destination – but now designated as the alternate – being only subject to a factor of 1.43 (when applied to the *required* landing distance) as appropriate to an alternate, instead of 1.67 as appropriate to a destination. Of course, if the real alternate also exerts a landing distance limit, then there may be little point in trying this ploy. (At the time of writing it is proposed – in the UK at least – to abolish this differential.)

'Payload Loss' System

Another way of calculating payloads for quotations in a hurry is to use the 'payload loss' method. In this method, every aircraft has its own maximum available payload worked out by the simple process of subtracting each APS weight for each configuration from the Maximum Zero Fuel Weight. Thus, using the leading figures from the foregoing examples, we have MZFW 69 000 kg. Taking the APS weight quoted for the 150 seats, one class layout we simply subtract this from the MZFW and the result is the maximum payload available for that configuration and APS. We can also establish, for each aircraft type, the maximum fuel load at brake release when the aircraft is at MZFW. This is simply MTWA–MZFW; in our example we subtract MZFW 69000 from MTWA 80000 and thus obtain the maximum fuel load that can be carried (from brake release), namely 11000 kg.

Example 1

> MSF 12000 kg = 1000 kg over max. fuel for max. payload. The payload loss for all aircraft of the same type in the same configuration is therefore 1000 kg. The limiting factor is minimum sector fuel.

Example 2

	RTOW	78 000 kg	
less	MSF	10 500 kg	(at brake release point)
	ZFW	67 500 kg	
	MZFW	69 000 kg.	ZFW is therefore 1500 kg less than maximum.

> Payload loss for all aircraft of the same type in the same configuration is therefore 1500 kg. The limiting factor is regulated takeoff weight.

Notes

As explained above, the maximum *possible* payload is simply the MZFW less the individual APS weight. MZFW is a structural limit, and is based on the maximum weight of the fuselage that the wing structure can sustain without some counter-balancing weight in the wings – i.e. fuel in the wing tanks.

This consideration can be well illustrated by taking the case of the Viscount turbo-prop aircraft which, although old, is still in service at the time of writing. This concerns a case where a fluid other than fuel affected the ZFW. This fluid was, in fact water-methanol, and its function was to restore takeoff power when the aerodrome was at a high altitude, or at a high temperature, or both, depending upon the version of the engine

(Rolls-Royce Dart) fitted. By injecting water-methanol fluid into the eye of the engine compressor, power could be fully, or partially restored for takeoff, the extent depending upon the rating of the engine in terms of temperature and altitude. In most other aircraft types where water-methanol, or other fluid – e.g. demineralised water – was used for this purpose the fluid was carried in fuselage tanks. In the Viscount the water-methanol tanks were carried in the inner wing, close to its root. It was therefore possible to specify a proportion of the fluid that could be allocated to the wing structure and thus to consider it as being fuel, for ZFW considerations.

Ideally, of course, the MTWA, MLWA, and MZFW should be the same, to allow for complete flexibility and maximum payload possibilities. But there is no merit in scheduling weights that cannot, in practice, be used and the only gain being that of use of a single value for memory retention. In our illustration aircraft, for example, with a MTWA of 80 000 kg a max. landing weight having the same value would be useless; the realistic MLW *must* be lower to account for the fuel burnoff on each flight. Likewise, were the MZFW to be the same as the MTWA and the aircraft loaded thus, then only the taxy-out fuel could be loaded. But what is useful to an operator is to have these various limiting weights as close together as possible, so that the aircraft can be used over a very short sector with a low burnoff, thus calling for a relatively small difference between MTWA and MLWA, and a MZFW close to the MLWA, so that only the reserve fuel required need come between the two. Such a pattern allows for the carriage of the maximum possible payload over short sectors, reducing according to stage length as the MSF required increases. Such an aircraft could be used on short, high density, sectors and also higher, medium or even long stages. And all without practical payload loss; the shorter stages would probably offer a volumetrically-limited payload, which would still be the maximum practicable usable, in fact. (A volumetrically-limited payload is one where the actual *weight* available cannot be utilised due to fuselage volume considerations, or number of seats. Only if there was a cargo of, say, gold bars could the highest permissible payload be carried.)

Note
Where the MTWA = MLWA could be useful, would be in the case of an emergency just after takeoff and necessitating a landing as soon as possible. If there is any gap between these two weights it would be necessary to jettison fuel so as to bring the weight of the aircraft down to the MLWA. And fuel is a very costly item. Even an engine failure in a multi-engined aircraft, while not normally being hazardous, would

indicate landing back at the same airport if this happened to be the operator's base. But, with a three or four engined aircraft, and under the circumstances being considered, it could be preferable to continue the flight if operationally acceptable, e.g. if the *destination* were to be the operator's base. The preferred course of action would always depend upon the actual circumstances, of course. A high MLWA eases the decision-making process.

So much for the more mundane and basic requirements for route planning. In the following chapters we deal with the more technical aspects of this very important subject.

3: An Example of a Worked Route Analysis

In the opening chapters we have covered the functions that may be deemed to be route planning activities in general. There may be other functions, according to an operator's own philosophy and requirements. However, the foregoing is felt to give the reader a good idea of the normal *raison d'être* for a route planning department in a medium to large airline. We now turn to a typical request, and response situation, in which route planning receives a request from the commercial department for a single charter, for passengers only, on a short stage. Detailed data is not yet required, and such information will be dealt with later in this book. All that is needed now is a block time, and fuel and payload availability on a stated stage. The example aircraft is based on a contemporary turbojet, type 'Y'.

We will analyse a route that is both well-known and well-used, but we will assume that the airline is new and has not yet operated this particular route. For our example we will assume that the stage in question is London (Gatwick) to Brussels. As the great majority of takeoffs from Gatwick are made using Runway 26L, due to the prevailing wind, we will assume a departure using this runway. The date is 21 January, and the time is 1300 GMT. The takeoff run available (TORA) is 10 164 ft, the accelerate stop distance available (ASDA) is 10 364 ft, and the takeoff distance available (TODA) is 10 663 ft. The elevation amsl is 202 ft, and there is negligible slope. The landing distance available is 9288 ft. (A full description of the implications of these values is contained in the companion to this book entitled *Handbook of Aircraft Performance*, by the same author.)

The expected takeoff temperature is +9°C, with a surface wind expected that will yield a takeoff wind component of 10 kt headwind. The statistical en route wind for flight level 250 is: 50% 28kt tailwind, 85% 5 kt headwind. The statistical en route temperature, in standard atmosphere, is ISA –5°C. The designated alternate is Amsterdam (Schiphol). Using the Dunsfold and Mayfield SID from Gatwick, via Dover, Koksy, and Nicky, plus a westerly standard arrival (STAR) at Brussels, we find that the stage distance is 225 nm. The diversion to Amsterdam is 111 nm, and the en route wind is 10 kt headwind.

The commercial department have indicated that they believe that the type 'Y' aircraft is the one most suited to their requirements, and accordingly the analysis will be based on this. As mentioned, a contemporary type has been used as a basis for the aircraft type 'Y', but the figures quoted do not refer to any specific aircraft type, and should not be interpreted as doing so.

First, the leading particulars of the aircraft. The limiting airworthiness weights are:

MTWA	33 000 kg	
MLWA	30 000 kg	
MZFW	27 000 kg	
APS	18 890 kg	(not an airworthiness weight)

Knowing the aircraft's runway requirements, and the airport runway length and elevation amsl, it is already known that there will be no weight limitations due to these values. We can therefore work out our example backwards from the MLWA, and using rapid planning tables or data.

The distance involved merits 'high speed cruise' on overall economy grounds, and we find that, using the appropriate table for the en route temperature deviation ISA –5°C the flight time from Gatwick to Brussels is 46 minutes, at flight level 250. Adding a block allowance of 10 minutes taxy-out (including start-up and checks), and 5 minutes taxy-in, we have a block time of 1 hour 06 minutes. The burnoff, engines on to engines off, is 2356 kg; 5% contingency allowance is included based on the flight element. The burnoff from brake release is 2276 kg. The reserve fuel for a diversion to Amsterdam, following a missed approach, pluse 5% contingency, and including 30 minutes holding at Amsterdam at 1500 ft, is a total of 1931 kg. This, added to the route fuel figure of 2356 kg, gives us a ramp fuel figure of 4287 kg, and brake release MSF of 4207 kg (assuming that 80 kg is burned prior to brake release; the MTWA may be exceeded by the weight of the taxy-out fuel).

Using MLWA 30 000 kg and adding to this the burnoff from brake release 2276 kg we obtain a required TOW of 32 276 kg. This is less than MTWA and is therefore not limiting. If we deduct from the required TOW, the weight of the fuel load at brake release i.e. 4207 kg, we find that the ZFW is 28 069 kg. But the MZFW, which *must* be observed, is 27 000 kg, and it is to this figure that we must add the MSF (brake release) of 4207 kg, which gives a revised required TOW of 31 207 kg, limited by MZFW and MSF requirements. When MZFW limits, of course, the payload available is the maximum permissible. But in order that this payload may be calculated we must bring in one more weight, namely the APS weight. The maximum ZFW–limited payload is simply

MZFW–APS wt (i.e. 27 000 kg–18 890 kg) and this is 8110 kg. This figure allows for every seat to be filled by a male, in fact. And this payload should keep *any* commercial department happy! (It is assumed here that standard weights are used; a standard male weighs 75 kg, and to this is added a luggage allowance of 15 kg, making a passenger unit weight of 90 kg. The APS weight is based on an 85 seat layout, so that the seat-limited payload is 85 ×90 = 7650 kg.) There is thus 460 kg weight in hand; this can either be utilised to carry some freight, or extra fuel. The carriage of more fuel than is required can be economically valuable, if, for example, the fuel price to the airline at the airport of departure is lower than that at the destination. But it must be remembered that *any* extra weight can cost money to carry, in terms of very marginally increased flight time and burnoff. Guidance to flight staff is normally provided to enable them to judge the value of uplifting excess fuel for economy considerations.

So, having done their sums to a reasonably accurate degree (the actual conditions prevailing on the day of operation may well cause a different set of values, although not wildly so as the 85% case has been used) the route planners will advise the commercial department simply the following, as the answer to their enquiry.

Block time	1 hour 06 minutes
Block burnoff	2356 kg
Payload	8110 kg (maximum permissible by regulation)

4:   Flight Documentation – General

Introduction

The next few chapters of this book are concerned with an in-flight document that is often referred to as a 'route book'. As has been mentioned earlier, this is but one title for this important document. Its purpose is to collect, simplify, and present operating data that, while being vital and therefore necessary, does not lend itself easily to reference on the flight deck. It is therefore important that an operating crew is provided with easily-assimilable operating data that is to an acceptable level of accuracy in comparison with that data upon which it is based. In certain cases flight manual (AFM) data is presented in such a form that it *could* be risky – if not dangerous – to attempt to use this away from ground facilities. Precalculation and presentation in a simplified format is therefore a *must*.

WARNING

The documents upon which Part 2 is based are believed to be up-to-date at the time of writing. They are, however, subject to amendment, and this could mean that the relevant data contained herein is, as a result, out of date. Therefore this part must be regarded as being for illustration purposes only, and **under no circumstances** should any chart or data contained therein be used for any operational purposes. Further, in addition to the 'dating' process, there is another hazard source, namely, that because of editorial considerations it has been necessary to redraw certain charts to a different scale. These charts could therefore be subject to drawing errors, although every effort has been made to maintain a high level of accuracy.

We have dealt with the function of route planning, and the possible structure of a route planning organisation, in the previous chapters. It must be clearly understood that both function and structure of such an organisation can, and probably will, vary from airline to airline. But the required 'end product' will almost certainly not. The various elements that comprise the route planning team may well have other duties and only come together when circumstances dictate. However, unless we are considering a *very* small airline, we can be reasonably certain that some specialist unit will be responsible for the production of time, fuel, and payload values on request. Such a unit will normally also be responsible for the production of flight documentation for use on the flight deck.

To be fully clear as to what is meant by the expression 'flight documentation' we should make ourselves aware of what documents are required to be available to the flight crew during the flight planning process and when airborne. First, there is the aircraft flight manual (AFM), and this document forms part of the Certificate of Airworthiness. Some countries, e.g. the UK, do not require the AFM to be carried, as long as certain specific sections – limitations and emergencies – are included in an operator's flight documentation (UK ANO Schedule 12 refers). It contains data concerning the mandatory limitations of the aircraft e.g. weights and speeds, and its certificated performance. The latter includes takeoff and landing distances, climb after takeoff, and certain instrument calibration and correction values, such as the airspeed indicator (ASI) and the altimeter. It does not contain any information regarding en route speeds (other than where these are limiting) or fuel consumption. It is often, due to necessity, a very complex document and is not intended for frequent aircrew reference in flight.

Then there is the airline's 'operations manual', and this is variable in layout. It usually comprises a number of separate volumes, and is based upon the actual operation of a company's aircraft under national requirements and international agreements, viz. adherence to the operations recommendations issued by the International Civil Aviation Organisation (ICAO). Reference should be made to ICAO Publication Doc. 8168 PANS-OPS/611, Volumes 1 and 2 for details regarding these recommendations. In the UK minimum acceptable requirements are also contained in CAA publication CAP 360 (national requirements).

A typical operations manual breakdown is where one volume defines the airline's structure, policies, chain of responsibilities, and items of a general nature, and a second volume, specific to a particular aircraft type, covers technical details of the aircraft and its systems. ('Pilot's Notes' is one expression that covers this volume, although this is not now in general usage.) Then there is the chosen 'route flight guide' e.g. Aerad or Jeppesen, and finally the volume that mostly concerns us in the context of this book. This volume contains data in simplified form concerning the operation of a particular aircraft type, and is often referred to as the 'route book'.

The Route Book
The primary purpose of the route book – and this term will be used

throughout this book – is to provide pilots and ground operations staff with simplified data that can be read easily and with the minimum of calculation activity. It may be said that it comprises precalculated flight data, from the flight manual, from cruise control data provided by the aircraft's manufacturer (i.e. longish-method flight planning data), from data contained in the flight guide, and from the technical manual. A typical route book will be divided into parts, or chapters, thus:

Part 1   Limitations and general (for the aircraft type).
Part 2   Precalculated takeoff data (from flight manual).
          Precalculated landing data (from the flight manual).
Part 3   Flight planning data for sector fuel and time (from cruise control data).
          Route details for each stage that the airline operates. This may be given in the form of a separate page for each sector or stage, with information on minimum safe altitudes, distances and magnetic tracks (Tr (M)) together with radio aids along the route and their frequencies and identifications, standard wind component used, alternate, and MSF. The data may also be issued in the form of a prepared log (plog), or may even take the latter form alone, without any page in the route book. If this is the case the plog must be declared to form part of the operations manual.
Part 4   Simplified aerodrome operating minima (AOM) tables (but these may be contained in the flight guide for each airport involved).
          Where an operator elects to prepare his own AOM, full values of specific AOM for each airport, runway and radio aid, must be tabulated. But in any event, the basic means of calculating acceptable AOM by the operating crew must be provided to allow for the rare, but possible, contingency whereby a crew may find that it has to operate a stage for which data has not been provided by route planning (e.g. when away from base).
Part 5   Emergency procedures (from the AFM and also the technical manual).

It is usually a good idea – at least in the opinion of the author – for each Part to be printed on different coloured paper, for ease of reference. Thus, Part 1 could be on white paper, Part 2 on green, Part 3 on blue, Part 4 on yellow, and Part 5 on pink. Or coloured divider cards could be used between each part as an alternative method. The route book is an important document and it well repays careful thought when deciding on its layout. In today's modern environment, even though flight planning is frequently done by computers, a flight crew still has to file anticipated flight time, sector fuel (and thereby, endurance) and fuel uplift, for the forecast conditions. More often than not this may need to be done during a short turnround – e.g. 1 hour. And until the fuel load is known, and the payload is notified, the required takeoff weight of the aircraft is not known. It must be possible for the crew or, if such a luxury is available, the flight despatcher, to calculate easily and simply the significant takeoff speeds V_1, V_r, V_2 and so on. And it is *not* reasonable to expect either the flight crew or the despatcher to have to use the AFM or the cruise control data to achieve this objective. For one thing there would probably not be time. The use of a computer helps, of course, but these revered devices have been known to fail, and thus all concerned must know both how to calculate the foregoing values and have time to do this. Therefore, access to a route book, when necessary, is essential.

Fuel Price Index
Another useful item to have available to all line flight crews is a table showing the contracted prices that the airline has negotiated for fuel at each regular stop, so that the uplifting of extra fuel, weight permitting, for economic reasons can be practised. But in the interests of commercial confidentiality it is best that the out-station prices are expressed as a percentage over or under base price e.g. fuel price index at Gatwick (base) is 100. Tenerife may be given as 110, which indicates that it is 10% more expensive and therefore nothing is to be gained commercially by uplifting more fuel than is operationally required, due to its weight.

Route Details and Minimum Sector Fuel
The actual layout of the route book must, inevitably, depend upon the size and activity of the airline concerned. A solely charter operator, for example, will need to spell out its fuel and AOM policies, and then provide written data for each flight through the medium of the flight brief (which will be gone into later in more detail). This is because a large number of destinations may be involved annually, often with a fairly low frequency of visit, so that the inclusion of specific data for each stage could well make the route book very bulky and unwieldly. Instead, the company's route, fuel, and AOM policy will be given, and each flight, or series of flights, will then have specific data provided by means of the flight brief accompanied by a plog, or plogs. Each flight brief and plog should preferably carry a note to the effect that it forms part of the company operations manual.

A predominantly scheduled service airline will normally spell out each route in some detail by means of specific pages in the route book section of the operations manual, although the route navigational data may be issued in the form of plogs. A 'library' of these may be kept in the route planning office, or at flight operations. In some cases, an operator may prefer to detail this information by means of a specific page or pages, in the route book, including times between waypoints, fuel burnoff expected for each leg of the journey for a statistical route wind component – normally the 85% value, minimum safe altitude (MSA), and details of the destination to alternate in a similar format. The fuel required for the departure to destination stage plus the destination to alternate stages is then totalled and a contingency allowance is added – normally 5%. The laid down holding allowance, plus the taxy-in allowance is then added, and the grand total fuel figure is declared to be the minimum sector fuel. Normally no flight may be allowed to depart with *less than* MSF, but it may be allowed to carry excess fuel. If the en route forecast conditions so demand, the *required* fuel for these conditions may exceed MSF. In such cases the former takes precedence, of course.

Each scheduled stage may be listed with a MSF value, but with the en route data contained in plogs to correspond. Where no plogs are issued, the crew or despatcher must transfer the route details from the route book to a navigation log prior to each flight. The advantage of the plog method is, in the author's opinion, that it keeps the route book thinner.

En Route Alternates

An en route alternate may be designated if necessary, subject to a number of conditions. The main reason for using such an alternate is that it permits the carriage of less than the full contingency fuel and therefore this weight saving can be made up by payload. It is therefore a practice that should only be resorted to in those circumstances where the RTOW is limiting, and, as mentioned above, special conditions must be met when it is used.

The main conditions for reducing the contingency fuel carried, when using an en route alternate, are:

(1) The fuel carried may not be reduced by more than the contingency fuel requirement based on the departure to en route alternate, (ERA).
(2) A technical refuelling stop *must* be made at the ERA unless the actual quantity of fuel available from the ERA complies in full with the MSF requirement stated earlier – in other words the aircraft must arrive at the destination missed approach point with the full reserve fuel laid down, and, in addition, must carry the 5% contingency on the ERA to destination.

(3) There must be forecast meteorological conditions for the ETA at the ERA that are at, or above, the company's AOM for that airport, at the time of departure.

Long Distance Routes

So far the route details methodology discussed refers to those appropriate to short-to-medium haul overland stages, usually between waypoints marked by a radio navigational aid, such as a VOR or NDB. Very often these routes follow controlled airspace, in the form of airways, or in other, but uncontrolled, legs known as advisory routes or ADRs along which advice, but not control, is available from ATC (air traffic control). Flights across long ocean stages have no waypoints, and each individual flight is routed according to the meteorological pressure pattern current for the expected departure time. Clearly it is not possible for route details to be laid down, and every flight track will almost certainly be somewhat different to the others that have been operated or will be operated in the future. For each flight a long oceanic crossing will have an optimum routing, and there will be great competition between various operators at, or near, the departure time to fly this geographical route. But minimum separation standards between aircraft must take precedence, and therefore each flight will be allocated a routing based on this requirement by oceanic ATC. Normally ATC will try to allocate routings as close as is possible to the optimum. With few en route navaids, the various aircraft *must* carry an approved on-board navigational system e.g. Omega or Inertial, or else a crew member who is qualified as a specialist navigator. It is normal practice, these days, for the actual route to be selected by a computer on the ground, before filing the flight plan.

In such circumstances there is very little flight data that can be provided by means of the route book, other than diversion routing to a designated alternate which (except in the case of an island destination) may well be provided with navaid-identified waypoints. In the case of an oceanic island destination, however, there may not be a suitable alternate within an accepted range. In such cases no alternate is designated, and the reserve fuel is based on a specified extended holding time – normally at least two hours.

This chapter has attempted to outline, without going into too much detail, the main items of in-flight documentation that are required by an airline. It must be emphasised, though, that neither the layout or scope are fixed by any form of legislation. The type of documentation required can vary from operator to operator, and from state to state. Even the type of operation and aircraft can call for a different format and contents. For example, as mentioned earlier, in the UK it is not strictly necessary for the flight manual to be carried, subject to the necessary information contained therein being made available through the airline's operations manual, which would appear to imply that the AFM contents must be transferred to the operations manual. Naturally, this would not make much sense, and in practice the extent of the AFM information that may be so transferred is negotiated with the appropriate regulatory authority i.e. in the UK, the Civil Aviation Authority. In the opinion of the author it is better to carry the AFM as there will almost certainly come an occasion when circumstances arise that were never allowed for and reference to the AFM may become imperative.

In the remainder of this part of the book we will examine the contents of the route book (as listed in this chapter) in greater detail, devoting a chapter to each of the route book parts where applicable. Please bear in mind that we are using *examples* of the type of requirements. In the case of limitations, for example, these can vary between aircraft of a similar type and variant, depending on the types of equipment fitted, including the engines. Even these can be identical for performance considerations but can have variations in their ancillaries that alter their limitations.

5: General and Limitations

There exists a large number of significant values and information in the AFM to which ready recourse is highly desirable. The AFM can be a very ponderous tome and trying to find a certain value can be both exasperating and time consuming. The real place for the AFM is in its allocated stowage on board (ready for reference when *really* required), and in route planning (where it is often required, and where space and instruments to enable it to be read accurately, exist). For example, how often does a UK operator need to check the minimum start-up outside air temperature (OAT), while the opposite may apply to, say, a Scandinavian airline? But certain leading numbers, values and data, are often needed for reference and these can be made into the contents of a chapter in the route book (or whatever title this volume, or part of the operations manual, may carry). This chapter will give a few examples of the sort of data that one might expect to find in an airline's route book. Once again, for illustration purposes, the BAe 146-100 has been used. It should be pointed out that simplification, in the interests of clarity, has been used, both in this chapter and throughout this part. For example, the ISA conversion chart (Figure 5.1) has been drawn in 5° intervals instead of the more usual 1° separation. Likewise, the takeoff weight, altitude and temperature (WAT) chart is only drawn for 24° flaps, and the 18° and 30° correction grid has been omitted (Fig. 5.2). In fact, in cases such as this, where there are a number of takeoff flap settings, it is, in the opinion of the author, probably better to provide a route book chart for each setting. As regards infinitely variable flaps this must appear as a plot against runway effects.

A word of caution. Certification limitations do not necessarily apply only to a specific *aircraft* type, or even variant of a type. Although each type and variant thereof, will have specified limitations, even certain installed equipment may well be subject to limitations of one sort or another, *and these may well affect the whole aircraft*. Normally items such as engines will be included as a variant to the aircraft's type, and therefore subject to that aircraft's overall limitations directly. But items such as the auxiliary power unit (APU), or flight management system (FMS) may have certain characteristics that result in the system limitation being more stringent than the aircraft's own. Naturally, unless there are very sound reasons for doing otherwise, a customer will try to avoid specifying on-

ISA – ALT./TEMP.

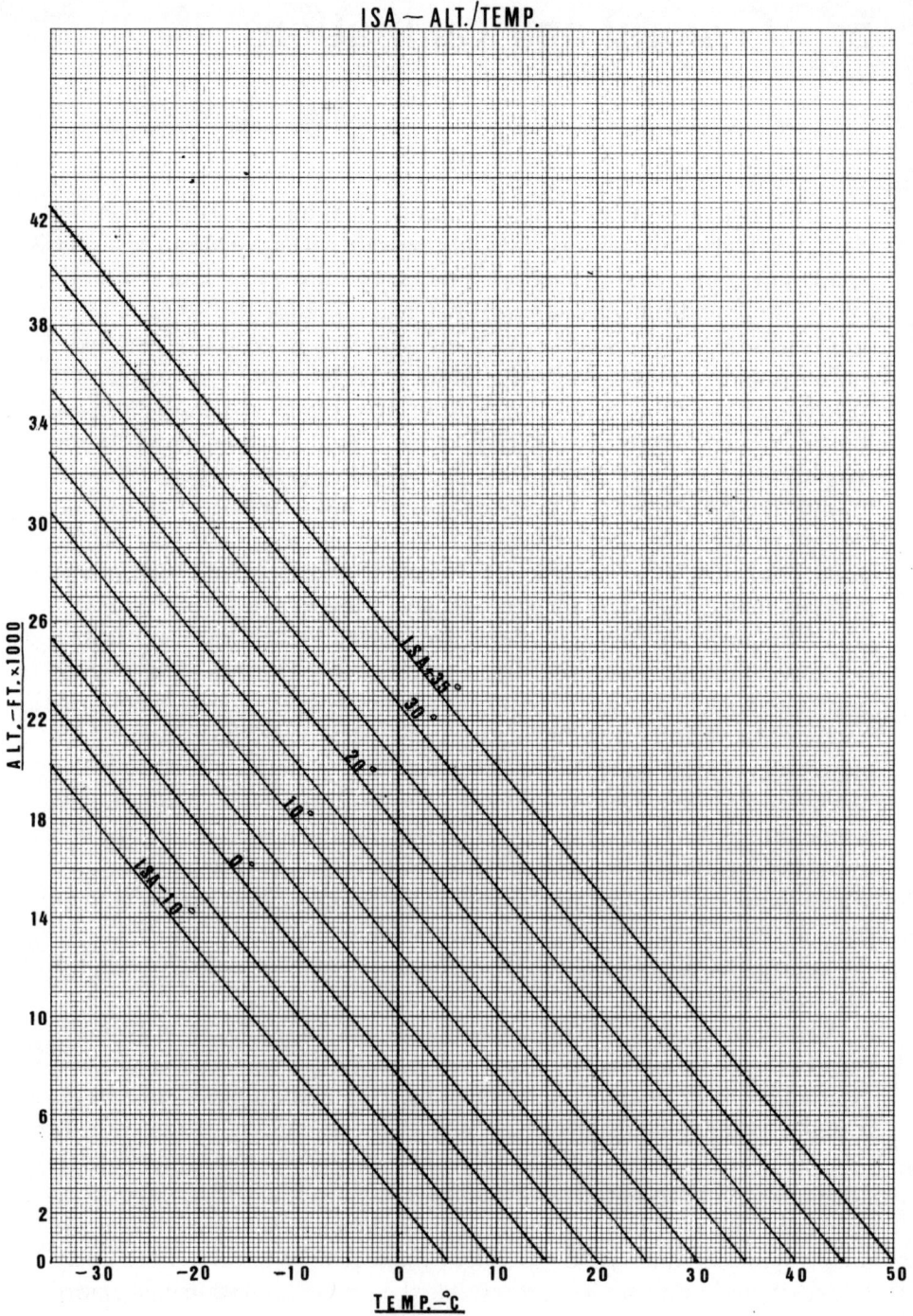

Fig. 5.1 Variation of International Standard Atmosphere with ambient temperature and altitude.

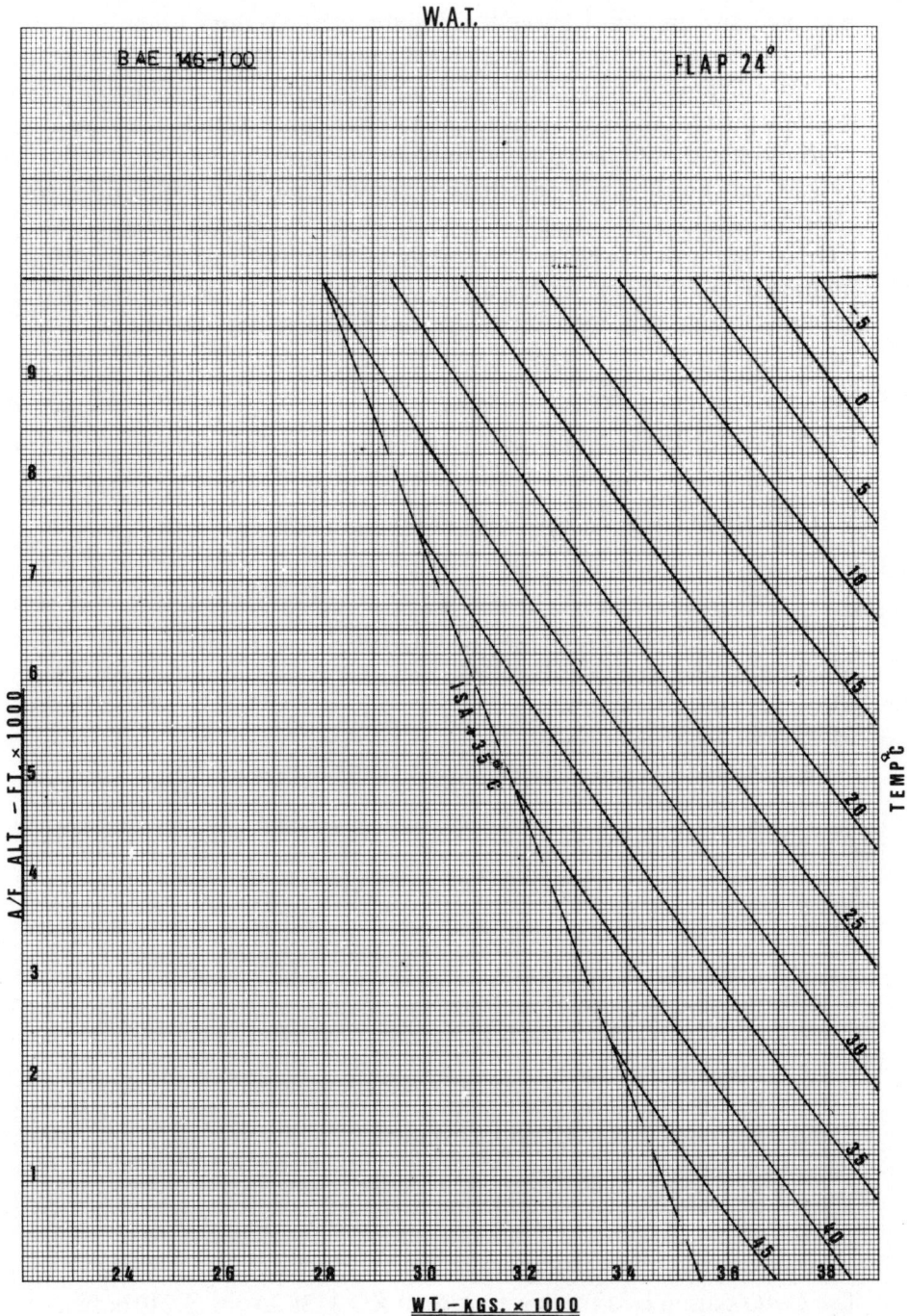

Fig. 5.2 The weight-altitude-temperature (WAT) limitation for the BAe 146, illustrated for simplicity using the 24° flaps case. In the flight manual this chart provides for corrections when using 18° and 30° flaps. A similar layout is used for the landing WAT case.

board equipment that is more limiting than the aircraft's limits.

So, let us now take a look at what might be described as a typical general and limitations section of a route book. It is only an example, and does *not* purport to be a complete chapter but rather a guide as to the sort of thing that should be expected to be found in such a chapter.

It is convenient to divide the limitations chapter up into different classified sections in a logical sequence. Thus, first comes 'weights', and so on. What now follows is a typical method for portraying general and limiting data for the example chosen, within the caveat contained above. Please note that in some variants the published values may differ.

BAe 146–100 (ALF-502R-3 engines)
(*Note power-plant qualification*)

1. Limitations (as at 22 February, 1989)

1.1 Weights

(a) Maximum Structural Weights Certificated

Maximum total weight authorised:	38 329 kg
Maximum takeoff weight:	38 102 kg
Maximum zero fuel weight:	31 071 kg
(The maximum weight of the loaded aircraft minus usable fuel load)	
Maximum landing weight:	35 153 kg

(b) Loading
Maximum floor loading
Forward baggage compartment:

max. permissible total floor load:	1170 kg
max. permissible floor load, kg/sq.m:	366

Aft baggage compartment:

max. permissible total floor load:	1097 kg
max. permissible floor load kg/sq.m:	366

Passenger cabin:

max. permissible floor load kg/sq.m:	195

(c) Centre of Gravity (CofG)
The CofG datum is at fuselage station AXO 1138.25 cm (1.210 m fwd of the reference point indicated at the rear end of the undercarriage wheel well pressure floor).

1.2 Speeds

(a) Structural Airspeed Limitations
Max. operating speed (V_{MO}):   300 kt IAS
Max. operating Mach no. (MMO)   0.7 indicated
Max. normal operating Mach no. (M_{NO})

(b) Max. Flaps Extension Speed (V_{FE})
Retracted: For *en route* and normal holding V_{MO}
18°: for takeoff and approach,   200 kt IAS
18°: low speed holding,   165 kt IAS
24°: for takeoff and approach climb,   170 kt IAS
30°: for takeoff,   160 kt IAS
33°: landing,   140 kt IAS

(c) Max. Landing Gear Operation (V_{LO} and V_{LE})
(Including flight with gear extended)   205 kt IAS

(d) Max. Speed for Acceptable Bird Strike
Not to be exceeded *below* 8000 ft   205 kt IAS

(e) Max. Manoeuvring Speed (V_A)
Flaps retracted,   205 kt IAS
Flaps 18°,   165 kt IAS

(f) Max. Autopilot Engagement Speed
Within the range V_{MO}/M_{MO} minus 5 kt IAS

1.3 Altitude and Temperature

(a) The Max. Takeoff Weight for Altitude and Temperature, (WAT)
Flaps 24° is presented at Fig. 5.2. (A similar presentation can be adopted for flaps 18° and 30°.)

(b) The Max. Landing Weight for Altitude and Temperature
(Similar in layout to Fig. 5.2. Normal landing flaps: 33°.)

(c) Max. Operating Altitude
Takeoff and landing: 6000 ft
30 000 ft (flaps retracted)
14 000 ft (flaps operating)
20 000 ft (undercarriage operating)

(d) Max. Operating Temperature
Takeoff, en route, or landing: ISA + 40°C *or* 50°C ambient

(e) Min. Ground Temperature Soak: ISA –40°C

2. Runways

(a) The aircraft may only be operated from hard paved surfaces.

(b) The Maximum Crosswind Component

Takeoff:	35 kt
Landing:	35 kt

  (with brake action GOOD)

(c) Runway Contamination (Normal Operations)

Maximum depth of water or slush covering the runway:	3 mm ($\frac{1}{8}$″)
	(takeoff and landing)
Maximum depth of puddles for takeoff:	25 mm (1″)
Maximum length of puddles for takeoff:	6 m (20 ft)

(d) Operation from Precipitation Covered Runways

Maximum depth of continuous layer of water, slush, or wet snow:	12$\frac{1}{2}$ mm ($\frac{1}{2}$″)
Maximum depth of continuous layer of very dry snow:	50 mm (2″)

3. Fuel and Oil Specifications

(a) Approved Fuel Specification
Wide cut fuels only when BAe Mod. HCM 5005 ZA incorporated and when fuel specification is listed below:

   DERD 2482 (UK) –
   DERD 2494 (UK) – ASTM D 1655 Jet A1 (USA)
   DERD 2453 (UK) – MIL-T-83133 (JP8) (USA)
   DERD 2498 (UK) –
   DERD 2452 (UK) – MIL-T-5624 (JP5) (USA)

(b) Approved Oil Specification
Type 1 (MIL-L-7808)
(EXXON/ESSO/ENCO/Humble 2389. AVREX-S-Turbo 256 (Mobil):
Includes IDGs, Main engine starters, APU Gen. adapter gearbox.
Type 2 (MIL-L-23699)
EXXON/ESSO/ENCO/Humble 2380. Mobil Jet Oil II: NOT for IDGs, Main engine starters, APU Gen. adapter gearbox.

4. Miscellaneous

(a) Minimum Crew

Minimum operating crew:	2 pilots

(b) Maximum number of occupants
Including operating and cabin crew; *excluding* children in arms: 94

Note
The foregoing is not a complete list, and also may not apply to all BAe 146
variants. It is provided solely as an illustration as to the type of
information that can be usefully presented in this part of the route book.

6:   Airfield Performance

With ever-increasing complexity, performance and techniques, the calculation of takeoff and landing values becomes a potentially very long-winded business. All the necessary data is contained in the flight manual (AFM), but using this data is something that cannot be expected for day-to-day usage. By way of example, to 'read' the AFM charts may well require the use of a simple drawing instrument known as a french curve. To get the best out of AFM data it must be accurately 'read' and noted. Intricate plotting may well be necessary and it is simply not practicable for this process to be accomplished as part of the preflight departure procedures. Equally, though, it is essential that all concerned, including flight crews, know how to calculate takeoff and landing data from the AFM in case an unexpected occasion should arise. For what is involved the reader is referred to *Handbook of Aircraft Performance*, published as a companion volume to this current work, by BSP Professional Books.

Format of AFM Data
There is no common format for the AFM yet, although there is a trend towards the general use of the 'web' chart for takeoff performance. In this form of presentation it is possible to enter a chart with the accelerate-stop distance that is available, (ASDA) and the takeoff distance available (TODA), and then to correct both values for the runway slope and surface wind reported. This process will yield an equivalent balanced field value, and a $V_1:V_r$ ratio (a balanced field is where the ASDA = TODA). Such a method allows for differing values of ASDA and TODA to be used, for example, where the TODA is greater than the ASDA and may be used for the initial climb. The difference between ASDA and TODA is known as clearway, and the availability of clearway can be a very important factor in permitting a higher takeoff weight than would be the case were the single value of balanced field to be the sole parameter. So the 'web' chart permits any clearway that may be available to be used to produce a value that is *equivalent* to a balanced field value. This is not to say that the value so obtained will equal the TODA, but it will probably be significantly more than either the ASDA *or* the balanced field length. As an illustration only, consider the case of a runway having an ASDA of 2000 m. (It may be assumed here that the TODA will also be 2000 m, and this is the classic balanced field available.) But, there may also be clearway

available, and if so this will be promulgated by the competent authority. This clearway may be added to the 2000 m to produce an increased TODA, e.g. 200 m clearway will then give a TODA of 2200 m. The interaction of the ASDA and the higher TODA will produce, through the use of the 'web' charts, a higher value of equivalent balanced field than is possible where the AFM only schedules balanced field data. (Example: MD80 aircraft AFM.)

In addition to the ASDA and TODA we also need to consider the takeoff run available (TORA), and the all-engines TOD performance – the latter in the case of aircraft having more than two engines; twins are never limited by the all-engines TOD required. The TORA may be less than the ASDA, and it is therefore necessary to establish that the value of TORA is sufficient to be non-limiting. A 'web' chart, similar to that used for ASDA v TODA is normally provided in the AFM, with TORA being substituted for TODA. It is used in exactly the same way, though, and produces a value of equivalent balanced field and $V_1:V_r$. Only in very rare cases will ASDA v. TORA be more limiting than ASDA v. TODA, but the check must be made. Incidentally, in some cases the equivalent balanced field value is simply referred to as the 'D' value, or 'D' in the case of ASDA v. TODA and 'R' in the case of ASDA v. TORA. The takeoff performance (runway) of the BAe 146 is presented in its AFM as 'D', be this a function of ASDA v TODA, ASDA v. TORA, or all-engines TOD required; *whichever is the lower value found is the limiting*. Fokker, taking the F28/4000 as an example, combines ASDA, TODA and TORA on one chart, using a 'web' format. But before the 'web' chart is entered, the TORA is augmented by any clearway available, up to a permitted maximum value of clearway, and after a further correction for surface wind and slope a value of 'reference' TOD is obtained. This is then associated with the ASDA (corrected for wind and slope) so as to yield a reference distance with $V_1:V_r$. (Incidentally, the F28 AFM allows for a check on the all-engines case, and yields a reference TOD using the same procedure as in the case of the TORA and clearway for the normal engine failure case, as described above.) The all-engines, and the engine-out reference TOD values thus obtained are then used to establish two takeoff weights, and lowest of the two being the limiting.

Regulated Takeoff Weights
The establishment of the RTOW in both examples cited above follows a similar trend. This is the addressing of a max. takeoff weight chart with the reference distance (in the case of the F28) or the 'D' value (using the lowest obtained, as described above) in the case of the BAe 146. The max. TOW or RTOW charts are, in most aircraft types, fairly similar in use, but

the layout may vary considerably. Essentially we have a value of runway available and we are seeking a value of RTOW from this. In essence the RTOW chart, irrespective of layout, provides a value of RTOW corrected for the effects of the pressure altitude of the runway and the ambient temperature that is being used, or is actually existing. (Statistical data can provide an appropriate value for planning purposes.) Then, having obtained a value for RTOW, a further chart is addressed using this RTOW and the V_1:V_r ratio so as to obtain the various takeoff significant speeds.

The use of 'D' values is, in fact, quite widespread in practice. It is the belief of the author that this form of presentation was first used in the British AFM for the Vickers Viscount, under BCARs. The FAR 25 AFM for the same type of aircraft was drawn so as to provide balanced field length values. Even now some quite modern types, particularly when originally certificated under FARs, still use the balanced field as being the yardstick, and thus it becomes virtually impossible to feed in any extra clearway that may be available. So the 'D' value presentation has two main characteristics of value. Firstly, it is *relatively* easy to work through (although, as has been mentioned, to obtain an accurate result special drawing aids must be used) and is more or less self-evident in its use. But it is still not really suitable for everyday use by the flight crews or despatchers. The second characteristic is that of a mainly commercial nature: any extra clearway may be taken into account when calculating RTOW, up to the maximum permitted under the applicable regulations, and this can result in increased RTOW values.

So far we have only considered two aircraft types, namely the BAe 146 (our main example aircraft) and the Fokker F28. Both of these are essentially in the 'small' category when considering transport-category aircraft, i.e. those exceeding 5700 kg MTWA. However, the former has four engines while the F28 is a twin. (The F28 has been further developed and is now known as the F100.) Both are European-built, but the data being used for the purpose of this book is drawn to joint airworthiness requirements (European JAR 25) in the case of the BAe 146 and to the US FAR 25 in the case of the F28. But we can refer also to a very much larger aircraft – in this case having three engines, and nearly five times the weight of the BAe 146 or nearly six times the weight of the F28. This large aircraft is used as the example type by the UK CAA for the purpose of training professional pilots in the use of the AFM for performance matters. Apart from the weight of this aircraft, however, the layout of the AFM charts is to a very similar format as the two smaller examples already quoted. In this example we find that we start off with TORA, add any clearway available, up to the maximum permitted (this latter value being drawn in to the chart, thus obviating the need for checking that no excess clearway is being used), and then correct it for runway slope and wind component. This results in a value of 'D' appropriate to all engines operating. We next take the ASDA and apply this to the slope and wind component correction grids, thereby arriving at an *equivalent* value of ASDA. Now we come to the familiar 'web' chart, but with a slight difference. We have already adjusted the ASDA, of course, so we use this value to enter the 'web', without any further correction. We also enter the chart with TORA, and correct this for any clearway (thus achieving a value of TODA). We now correct it for slope and wind component and move across the chart to intercept the ASDA line. As in the other two cases mentioned we arrive at a 'D' value and $V_1:V_r$. Max. RTOW is then discovered by addressing a further chart with 'D', pressure altitude, and outside air temperature (OAT). Using this weight, and the $V_1:V_r$ value already obtained, a further chart yields the speeds V_1, V_r, and V_2, including a check on V_{MCG}. Where appropriate a check can be made, by means of other charts, in all three cases to ensure that tyre speed limits and brake energy limits are not infringed. It should perhaps be mentioned that the 'specimen' aircraft used by the UK CAA training publication (CAP 385) is believed to be US built (but is not identified) and in the case under discussion is certificated to UK requirements.

In all cases, but varying according to certification standards, means of correcting for contaminated or icy runways is provided in the AFM concerned. It is essential that this data is borne in mind.

Precalculated Takeoff Performance Data

It has been indicated already that even the simplest AFM data presentation is not suited to everyday operational use. Instead, in the relatively unstressed confines of route planning, data may be accurately extracted from the AFM and then presented in the route book in such a format so as to require the minimum of interpretation. There are a number of devotees of graphical presentation, and others of tabulation. Probably a mixture of both is the best. Let us take just one method of dealing with RTOW to start with. We will use the BAe 146-100 as our example again.

The first requirement is to establish the 'D' values yielded by each runway in the airline's fixed route network. To achieve this the published values of TORA, TODA and ASDA are noted. For each runway threshold the elevation is noted, and by applying the difference to the length of the runway a percentage of the runway length is obtained, this being the *accepted* value of slope (see *Handbook of Aircraft Performance*). Provided with these values of TORA, TODA and ASDA, together with

Table 6.1

Airport and elevation	R/W	Distances in metres			%	"D" for Wind Component – kts and Ratio											
		ASDA	TODA	TORA	Slope	10T		0		10H		20H		30H		Limit
London (Gatwick)	26L	3233	3311	3159	0	2460	1.0	3220	1.0	3200+	1.0	3200+	1.0	3200+	1.0	nil
(202 ft) EGKK	26R	2565	2565	2565	0.04U	1860	1.0	2550	1.0	2700	0.98	2860	0.98	3010	0.99	nil
	08R	3233	3250	3098	0	2360	1.0	3200+	1.0	3200+	1.0	3200+	1.0	3200+	1.0	nil
	08L	2565	2565	2565	0.04D	1900	1.0	2575	0.98	2740	0.98	2880	0.98	2980	0.99	nil
London (Heathrow)	09L	3902	3902	3902	0.02U	2780	1.0	3200+	1.0	3200+	1.0	3200+	1.0	3200+	1.0	nil
(80ft) EGLL	09R	3658	3658	3658	0.02U	2740	1.0	3200+	1.0	3200+	1.0	3200+	1.0	3200+	1.0	nil
	27L	3658	3658	3658	0.02D	2760	1.0	3200+	1.0	3200+	1.0	3200+	1.0	3200+	1.0	nil
	27R	3902	3902	3902	0.02D	2800	1.0	3200+	1.0	3200+	1.0	3200+	1.0	3200+	1.0	nil
	05	2357	2357	2357	0	1720	1.0	2360	0.97	2500	0.97	2640	0.98	2900	0.98	nil
	23	2357	2357	2357	0	1720	1.0	2360	0.97	2500	0.97	2640	0.98	2900	0.98	nil

slope % up or down, we can now proceed to enter the 'web' charts and the all-engines chart and using various wind component values, tabulate the 'D' value of each runway (i.e. the most limiting values) under different wind components, within the limits contained in the AFM charts. Thus, referring to Table 6.1 we can see that under London (Gatwick) and London (Heathrow) we are given the elevation of the airport, the TORA/ TODA/ASDA, and the slope. In columns each 'D' value is then given for a different wind component at 10 knot intervals, viz. from 10 kt tail to 30 kt head, together with the V_1:V_r ratio. Taking London (Heathrow), Runway 09 left, we find that for a 10 kt tailwind component the 'D' value is 2780 and V_1:V_r is 1.0 (i.e. $V_1 = V_r$). For runway 05, with a 10 kt headwind component, 'D' = 2500 with V_1:V_r 0.97. It is permissible to interpolate between *adjacent* 'D' values to take account of different wind components e.g. for runway 05, and with a 5 kt headwind, 'D' for 0 W/C is 2360 and for 10 kt head is 2500. 'D' therefore increases at the rate of 14 per knot, so that 5 knots is worth 70 in 'D'. This may be either added to the 0 W/C figure for 'D' of 2360, or subtracted from the 10 kt 'D' value; the 5 kt W/C 'D' value is 2430, and there is no change in V_1:V_r. (Five knots is midway between the two 'D' values; were it not so interpolation would be best taken from the nearest 'D' value tabulated.)

Also provided on such a table are the ICAO location indicators (useful for filing the flight plan), the airport elevation amsl, and any limiting RTOW factor. For example, were we looking at a high elevation, in a high temperature, we might be WAT limited and not runway. In this case the 'D' values would be academic, although we would still need the V_1:V_r values. Linear interpolation between these ratios, as between 'D' values, is permissible, where applicable.

As regards slope, refer to Figure 6.1 (London, Heathrow). Clearly shown at the threshold to each runway is the elevation amsl for the threshold. Thus for R/W 09L we have an elevation of 80 ft at the takeoff end and 77 ft at the other end. The change in elevation for this runway is therefore 3 ft for a length of 3902 m (12 802 ft). The slope % is therefore

$$\frac{100}{12\,802} \times 3 = 0.2\%$$

And, as 77 ft is less than 80 ft, the slope must be downhill. It will also be seen, by reference to the top left-hand corner of the chart, that the airport elevation is 80 ft amsl.

The data extracted from Table 6.1 is, in effect, an *equivalent* value of a balanced field length together with an appropriate value of V_1:V_r. But what we now need is the RTOW. This may be achieved by either addressing a graph with the 'D' value, and after correcting for

Elev	Var	INS	See chart F1
80	5°W	RAMP	

(HEATHROW) LONDON AERODROME

START & CLEARANCE HEATHROW Delivery 121.7	PUSH/TAXI Ground 121.9	TAKE-OFF Tower 118.7 118.5	ATIS 115.1 133.075 113.75	D1	Ld
				06 APR 89	

EGLL

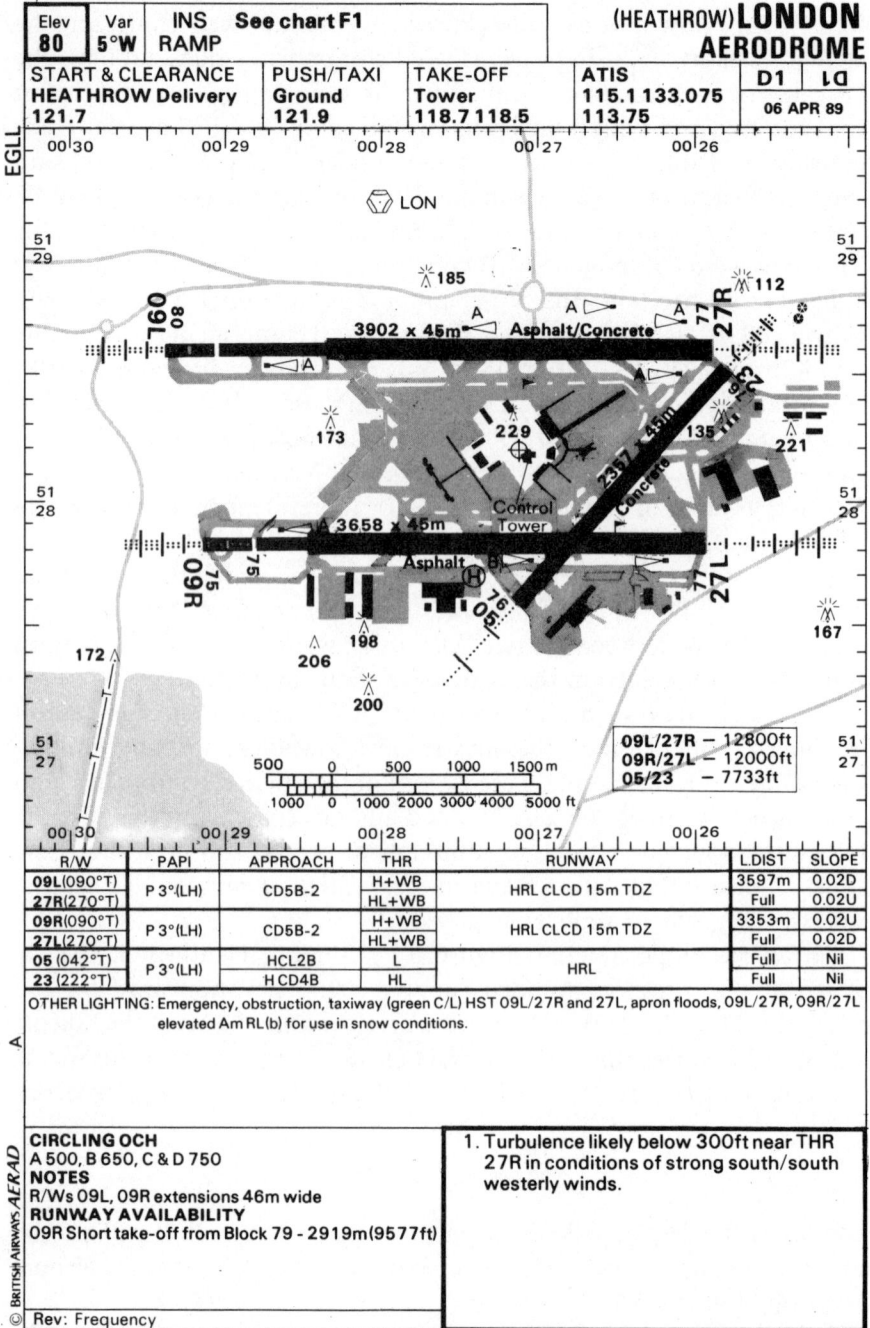

3902 x 45m — Asphalt/Concrete

185

112

173

229

135

221

3658 x 45m

Control Tower

Asphalt

198

206

200

172

167

09L/27R — 12800ft
09R/27L — 12000ft
05/23 — 7733ft

500   0   500   1000   1500 m
1000   0   1000 2000 3000 4000 5000 ft

R/W	PAPI	APPROACH	THR	RUNWAY		L.DIST	SLOPE
09L (090°T)	P 3°(LH)	CD5B-2	H+WB	HRL CLCD 15m TDZ		3597m	0.02D
27R (270°T)			HL+WB			Full	0.02U
09R (090°T)	P 3°(LH)	CD5B-2	H+WB	HRL CLCD 15m TDZ		3353m	0.02U
27L (270°T)			HL+WB			Full	0.02D
05 (042°T)	P 3°(LH)	HCL2B	L	HRL		Full	Nil
23 (222°T)		H CD4B	HL			Full	Nil

OTHER LIGHTING: Emergency, obstruction, taxiway (green C/L) HST 09L/27R and 27L, apron floods, 09L/27R, 09R/27L elevated Am RL(b) for use in snow conditions.

CIRCLING OCH
A 500, B 650, C & D 750
NOTES
R/Ws 09L, 09R extensions 46m wide
RUNWAY AVAILABILITY
09R Short take-off from Block 79 - 2919m (9577ft)

1. Turbulence likely below 300ft near THR 27R in conditions of strong south/south westerly winds.

BRITISH AIRWAYS AERAD

© Rev: Frequency

Fig. 6.1 The landing chart for London (Heathrow) airport. Runway lengths in metres, airport and threshold elevations in feet amsl, obstacles, circling minima (see Chapter 8), and radio frequencies are shown. (Courtesy of British Airways AERAD.)

Table 6.2 Conversion of 'D' into RTOW (kg) – MSL

D	ISA–10°C			Flaps	ISA		
	18°	24°	30°		18°	24°	30°
800	—	28 650	30 900		—	27 000	29 000
1000	30 900	32 650	34 200		29 300	32 000	32 700
1200	34 100	36 850	37 700		32 500	34 200	35 850
1400	37 000	38 000	38 000		35 300	37 600	38 000
1600	38 000	38 000	38 000		37 500	38 000	38 000
1800	38 000	38 000	38 000		38 000	38 000	38 000

D	ISA+10°C			Flaps	ISA+20°C		
	18°	24°	30°		18°	24°	30°
800	—	—	27 500		—	—	—
1000	27 500	29 500	31 000		—	27 100	28 850
1200	30 800	32 500	34 350		28 400	30 350	31 950
1400	33 250	35 000	36 500		31 300	32 850	34 200
1600	35 500	37 300	38 000		33 150	34 800	35 950
1800	37 300	38 000	38 000		34 750	36 650	37 900
2000	38 000	38 000	38 000		36 550	38 000	38 000
2200	38 000	38 000	38 000		38 000	38 000	38 000

D	ISA+30°C			Flaps	ISA+35°C		
	18°	24°	30°		18°	24°	30°
800	—	—	—		—	—	—
1000	—	—	27 000		—	—	—
1200	26 800	28 500	29 850		25 950	27 600	29 150
1400	29 200	30 850	32 150		28 550	30 000	31 250
1600	31 200	32 750	34 000		30 200	31 850	33 100
1800	32 950	34 500	35 800		31 950	33 600	34 800
2000	34 500	36 150	37 600		33 500	35 100	36 500
2200	35 900	37 600	38 000		34 900	36 650	38 000
2400	37 300	38 000	38 000		36 200	38 000	38 000
2600	38 000	38 000	38 000		37 400	38 000	38 000
2800	38 000	38 000	38 000		38 000	38 000	38 000

Note:
'D' values are given in 200 'D' Intervals. Interpolate linearly between adjoining tabulated 'D' values to obtain RTOW, e.g. ISA+10°C, flaps 18°, 'D' = 1750. RTOW for 'D' = 1800 is 37 300 kg, and for 'D' = 1600 is 35 500 kg. Difference in weight is 1800 kg, or 9 kg per metre 'D'. 1750 'D' is 50 m less than 'D' 1800, so RTOW is 37 300 – (50 × 9) = 36 850 kg. (18° flap assumed).

temperature and flap, arriving at a weight, *or* RTOWs may be tabulated against 'D' values, using the same corrections. One method of presenting this is as shown in Table 6.2, which is given in a somewhat simplified form, in the interests of clarity. In practice it would be preferable to list 'D' values for at least 100 'D' intervals and possibly 5° temperature intervals, although the note to the table gives a correction factor. Again, linear interpolation between two adjoining values of 'D' is permissible, as in the example cited above.

In this tabulation of RTOWs for 'D' values, each page is valid for a single airport altitude or elevation. It is normal practice to use this 'height' parameter, although it should strictly be pressure altitude. And one should expect to find a page for each 1000 ft difference, ranging from 1000 *below* msl (yes, there are a few such: Amsterdam (Schiphol) is one case, being 13 ft below msl) to 8000 ft or more amsl. This will call for 10 pages at 1000 ft intervals. Once again, linear interpolation between two adjacent elevations is permissible, as in the case of 'D', etc. So the sequence is: obtain the appropriate 'D' value for the runway in the reported wind component, and the ratio $V_1:V_r$. Take the nearest elevation page *below* airport elevation and extract the RTOWs for the flap setting and the nearest ISA temperature values below and above this elevation. Interpolate to obtain the RTOW for the appropriate temperature in ISA deviation. Repeat this for the next elevation, in exactly the same way. We now have two values of RTOW both of which are valid for the appropriate temperature in ISA deviation. But one weight is only good for the nearest 1000 ft tabulation *below* the actual elevation of the airport, and the other is good for the nearest 1000 ft page above. In other words, the RTOWs are taken from two immediately adjoining elevation pages. Once again we must interpolate between the two RTOW values thus found to establish the deviation across the appropriate 1000 ft range, within which lies the actual elevation. As a general rule a weight deviation per 100 ft in elevation is acceptable, unless the aircraft is altitude sensitive. In which case the altitude or elevation tables may need to be increased so as to bracket 500 ft intervals.

We have now obtained a value of takeoff weight that is appropriate to the runway, and the conditions currently obtaining. While the process may sound somewhat complicated, in practice it becomes quite rapid and straightforward. But we will need to obtain other values, these being the appropriate speeds that relate to the RTOW found. Both V_r and V_2 are directly related to the RTOW, (or to the actual TOW, if less than maximum), as appropriate for the ambient air temperature and airport altitude. A simple graphical chart can be included after Table 6.2 that enables these two speeds to be ascertained for the TOW applicable. We

36000 kg

TAKE-OFF FLAPS	18°	24°	30°
V_R	121	113	106
V_2	133	122	114
V_{FTO}	172	172	172
V_{ER}	178		
$V_{REF\,33}$	117		

ABNORMAL FLAPS	30°	24°	18°	0°
V_{REF}	122	128	138	178

37000 kg

TAKE-OFF FLAPS	18°	24°	30°
V_R	123	115	108
V_2	135	125	116
V_{FTO}	175	175	175
V_{ER}	180		
$V_{REF\,33}$	119		

ABNORMAL FLAPS	30°	24°	18°	0°
V_{REF}	124	130	140	181

38000 kg

TAKE-OFF FLAPS	18°	24°	30°
V_R	125	117	110
V_2	136	127	117
V_{FTO}	177	177	177
V_{ER}	182		
$V_{REF\,33}$	121		

ABNORMAL FLAPS	30°	24°	18°	0°
V_{REF}	126	132	142	183

CORRECTIONS

Temperature and altitude corrections to tabulated speeds

V_R AMB. TEMP. °C

PRESSURE ALTITUDE –ft	0	10	20	30	40	50
S.L.	−1	−1	−1	0	0	+2
2000	−1	−1	−1	0	0	+2
4000	−1	−1	0	+1	+2	+3
6000	−1	0	+1	+2	+2	+3

V_2 AMB. TEMP. °C

PRESSURE ALTITUDE –ft	0	10	20	30	40	50
S.L.	+1	+1	+1	0	0	−1
2000	+1	+1	+1	0	0	−1
4000	+1	+1	0	−1	−1	−2
6000	+1	0	−1	−1	−2	−2

Fig. 6.2 Tabulated takeoff and landing data for three specimen weights, for the BAe 146–100. Note small variation in speeds for 1000 kg weight changes – max. 3 kt. The abnormal flaps V_{REF} are for use when using flap settings other than the normal 33° for landing. The tables are for display, according to the weight on the co-pilot's panel. (Courtesy of British Aerospace.)

have already established the ratio $V_1:V_r$; knowing V_r for TOW to obtain V_1 we simply apply the ratio as a percentage of V_r. Thus, $V_1:V_r = 0.97$ is indicating that V_1 is 97% of V_r, e.g. if the TOW is 37000 kg and the V_r is 123 kt, the V_1 will be

$$\frac{123}{100} \times 97 = 119 \text{ kt}$$

But this is an arithmetical solution; normally one would set the V_r against the 100 mark on one's navigational slide rule and read off the value against 97.

A simple way of providing information on speeds V_r, V_2, and other speeds, at 1000 kg intervals is presented in Figure 6.2. This is issued by British Aerospace and relates to the BAe146-100 aircraft. Assuming flaps 18° we find, for example, that at 37 000 kg TOW, V_r is 123 kt, V_2 is 135 kt, V_{FTO} (final takeoff speed) is 175 kt, and the 'cleaned up' en route speed is 180 kt. The V_{REF} approach speed for 33° flaps (normal landing flaps) is 119 kt. This latter speed is, of course, only applicable to the approach and landing, as are the abnormal flaps V_{REF} speeds. Correction to V_r and V_1 speeds are also provided in Figure 6.2 to take into account the effects of ambient temperature and airport pressure altitude. Linear interpolation between adjoining 1000 kg values is permissible. In use, the booklet, opened at the correct page, slips into a holder on the co-pilot's panel.

Contaminated Runways

Once again, it must not be forgotten that all the foregoing is applicable to dry runways. Suitable corrections to all data as illustrated in the foregoing paragraphs must be provided to allow for wet or contaminated runways, and this will be provided, in 'raw' form, by means of the AFM. This correction can take the form of increasing the EMD requirement (ASD required), or a reduction in the value of V_1, or both. The V_1 thus obtained does not necessarily contain the same level of safety as it does for dry conditions and an element of risk is accepted over and above that for dry conditions. The AFM correction may take the form of graphical charts, or it may be a tabulated correction. Any such data should be processed from the AFM format, if necessary, so as to provide a correction to the precalculated dry runway data already provided.

Sometimes precalculated takeoff data cannot be presented quite so simply as in the case of the formats so far described. This is because, in order that certain TOW restrictions caused by factors other than the runway length can be alleviated, extra considerations have to be taken into account (see *Handbook of Aircraft Performance*). Several aircraft manufacturers now schedule additional data that is used to ease any second segment or WAT limitations, that arise due to a combination of airport elevation and OAT. The basic principle is to convert 'spare' or excess runway distance, into speed over and above that required normally. Thus, where V_2 (safety speed for takeoff) is normally $1.2V_S$

(stalling speed in the T/O configuration) any excess runway availability can be used to increase V_2, up to an approved amount according to the aircraft. For example, Fokker schedule data in the case of both the F27 and F28 that results in a significant easing of the WAT limit when V_2 or V_r (rotation speed) can be increased above the normal minimum speed 1.2 V_S and this also applies to the Lockheed L1011 TriStar. McDonnell Douglas, in their MD80 series (see Fig. 6.3), use a not dissimilar principle but invoke aerodynamic means. The MD80 has an infinitely variable T/O flap setting, instead of one or more discrete positions. Thus, knowing the TOW and OAT, the T/O flap is set to the optimum value for both the aircraft elevation and temperature under these conditions and also the runway length. Naturally, the lower the flap setting, expressed in degrees, the higher the value of V_S, and therefore V_r and V_2. So a speed increase is involved, but this is controlled by the mechanical action of flap setting.

The method of resolving these techniques, for the purposes of presenting max. TOWs and speeds, is normally to cross-plot WAT limits against runway limits, in such a way that, for actual values of weight,

Fig. 6.3   A McDonnell Douglas MD83 of Paramount Airways. (Courtesy of Bristol Airport.)

pressure, altitude, temperature, and runway both the max. permitted TOW can be found, with V_1, V_r and V_2, *or*, if the required TOW is less, then the appropriate takeoff speeds can be ascertained in a simple fashion. Where, for example, an infinitely variable flap setting is available the optimum flap position must also be found for the conditions.

Approach and Landing

Although there are comparatively few variables contained in the derivation of landing weight (and these are gone into in some detail in *Handbook of Aircraft Performance*, it is desirable that the landing performance requirements be presented clearly and simply. Firstly, it must be borne in mind that all landing distance requirements are based on ISA alone, *as appropriate to the pressure altitude of the airport*. In some cases, the landing distance required (LDR) varies according to whether the landing is to be made at the intended destination or at an alternate (see page 17). In the former case the *required* LD is the gross (i.e. unfactored) LD $\times$ 1.67, while for the latter the factor is 1.43. The approach is scheduled to be made at a given speed for the max. or required landing weight, and this is normally scheduled in the AFM. There can also be a threshold speed, V_{REF} or V_{AT}, and this is based on $V_S \times 1.3$.

As there is no temperature accountability for the approach and landing case (other than in the case of the limiting weight for altitude and temperature, which is an overriding limit concerned with a minimum climb capability while in the landing configuration to allow for a discontinued approach) it is not usual to find any LDA limits cropping up, except in exceptional cases. It is, therefore, normally sufficient to provide tabulated data giving the maximum landing weight permitted for all runways in the operator's network, including the alternates. In providing this data it must be clearly stated that cognisance has been taken of the pressure altitude, and also whether or not the listed weights are valid for destination, alternate, or both (it may be necessary to quote different weights in certain cases). A tabular presentation of V_{REF} and V_{AT_0} values may also be incorporated, for various values of weight. In Figure 6.4 the graphical format for V_{REF} against weight, with the normal landing flap 33°, applicable to the BAe 146-100 is provided for information.

Attention is drawn to the tabulated V_{REF} speeds for the BAe 146–100 example as presented in Fig. 6.2. It will be noted that the normal landing threshold speed is scheduled for flaps 33°, but values of 'abnormal' V_{REF} are also given for flap settings 30°, 24°, 18°, and 0°. The flap settings are used to cater for certain specific situations, such as emergency overweight landing, and 2 engines-out missed approach. When using these settings

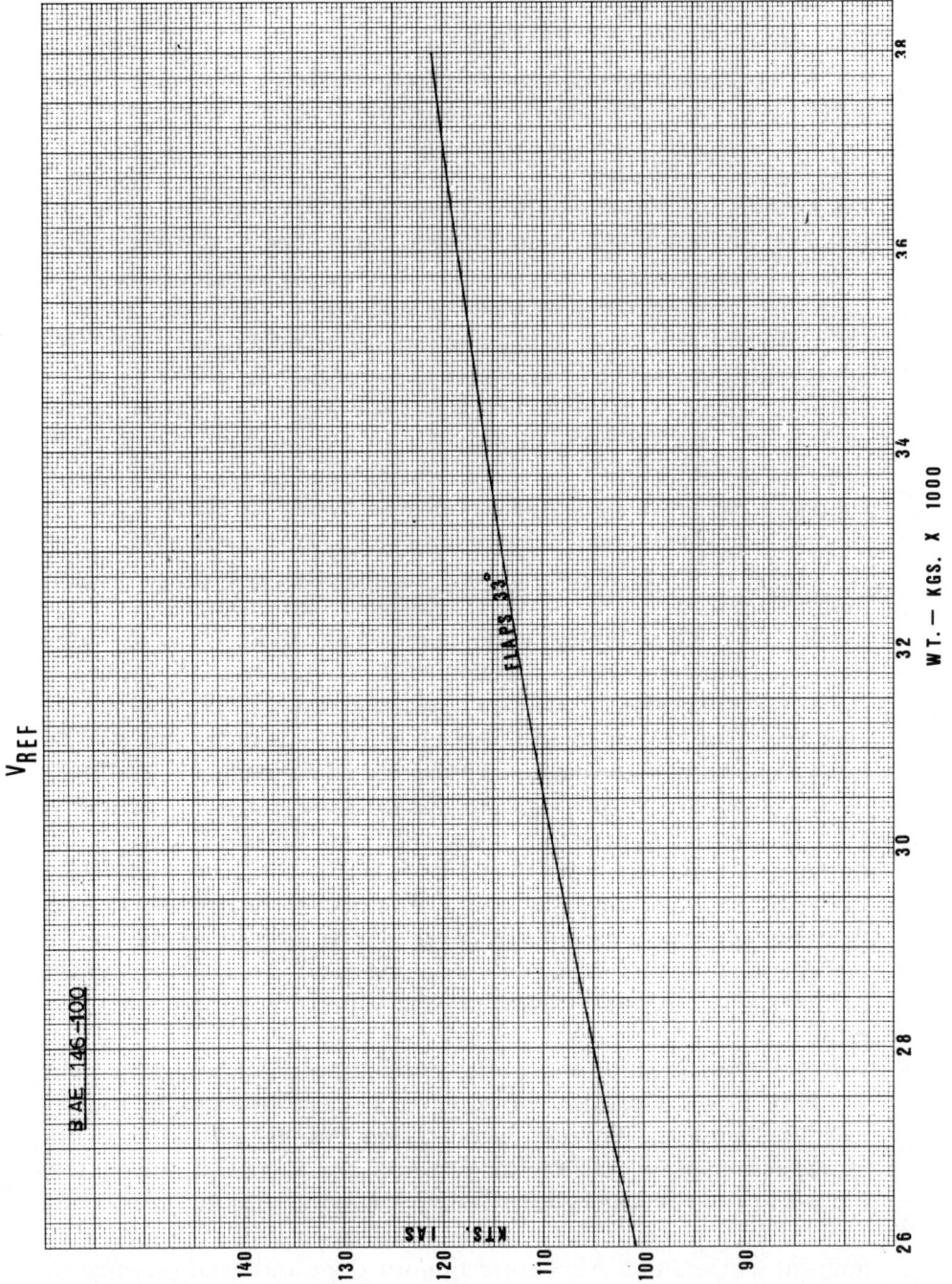

Fig. 6.4 The value of V_{REF} or V_{AT_0} (1.3 V_S) for the BAe 146–100 according to weight.

the landing distance available (LDA) must be reduced by 10% (30°), 20% (24°), 30% (18°) and 50% (0°). This reduction of LDA is quite arbitrary and is, in fact, carried out before entering the landing distance chart in the AFM.

Non-Runway Limitations

On occasion a situation may arise where the RTOW or LW is governed by values not resulting from the runway length. The most significant of these is weight-altitude-temperature or WAT. A WAT limit is usually a function of ambient temperature associated with altitude, and applies to the airport overall as a blanket limitation. It is normal to associate WAT limits with the so-called 'hot-and-high' airports, e.g. Mexico City or Nairobi. In such cases the RTOW is not affected by the runway length, and may be less than the value produced by the runway. This can mean that the 'D' values and $V_1:V_r$ are invalidated, and an entry in the 'limit' column of Table 6.1 should indicate this, i.e. WAT. To present corrected 'D' and ratio values first of all ascertain the RTOW from the WAT chart for a range of temperatures at, say, 5°C intervals and for these weights find the *required* 'D' and ratios. This simply means obtaining 'D' for weight and OAT and selecting the appropriate 'D' line from the 'web' chart. Draw in the ASDA and TODA (or TORA) corrected for wind and slope and note where these two lines intersect the 'D' line already identified. This process could well result in a low and high value of $V_1:V_r$ being identified. On the basis that WAT is the limiting factor, the listing of 'D' values is academic. Instead, list the ratio range for each W/C column, and add a note to the effect that the runway is not limiting through 'D' and that the V_1, V_r and V_2 should be obtained directly from the weight and ratio range. In some cases it may be the operator's policy that the highest possible value of V_1 should be used, in which case the higher end of the ratio range only should be listed. There are arguments for and against, using the highest possible value of V_1 appropriate to the ratio.

It is felt that an illustration to show how V_1 need not be a single value would be useful. Consider, for example, a situation where the runway length(s) is not the limiting factor. Taking the BAe 146-100 as our example, it may be that, for some reason, the RTOW is non-runway limited to (say) 36850 kg (or perhaps it may be that this is simply a *required* takeoff weight); the airport is at msl in ISA + 10°, i.e. 25°C ambient temperature. After correction for slope and wind the effective ASDA is 1800 m and the effective TODA is 1900 m. These two values combine to produce an *available* 'D' value of 1860, and a $V_1:V_r$ of 0.94. But, using the RTOW for 'D' tables (Table 6.2) we find that for the TOW required 36 850 kg, the 'D' value requirement is only 1750. If we leave the ASDA and TODA lines plotted in on the 'D' and V_1:V_r, and producing the values noted above, we can draw in a 'D' value line across the 'web' of 1750 (this being the required 'D' value) (see Fig. 6.5). It will be seen that this 'D' line intercepts the corrected TODA line at ratio 0.87 and the ASDA line at 0.96.

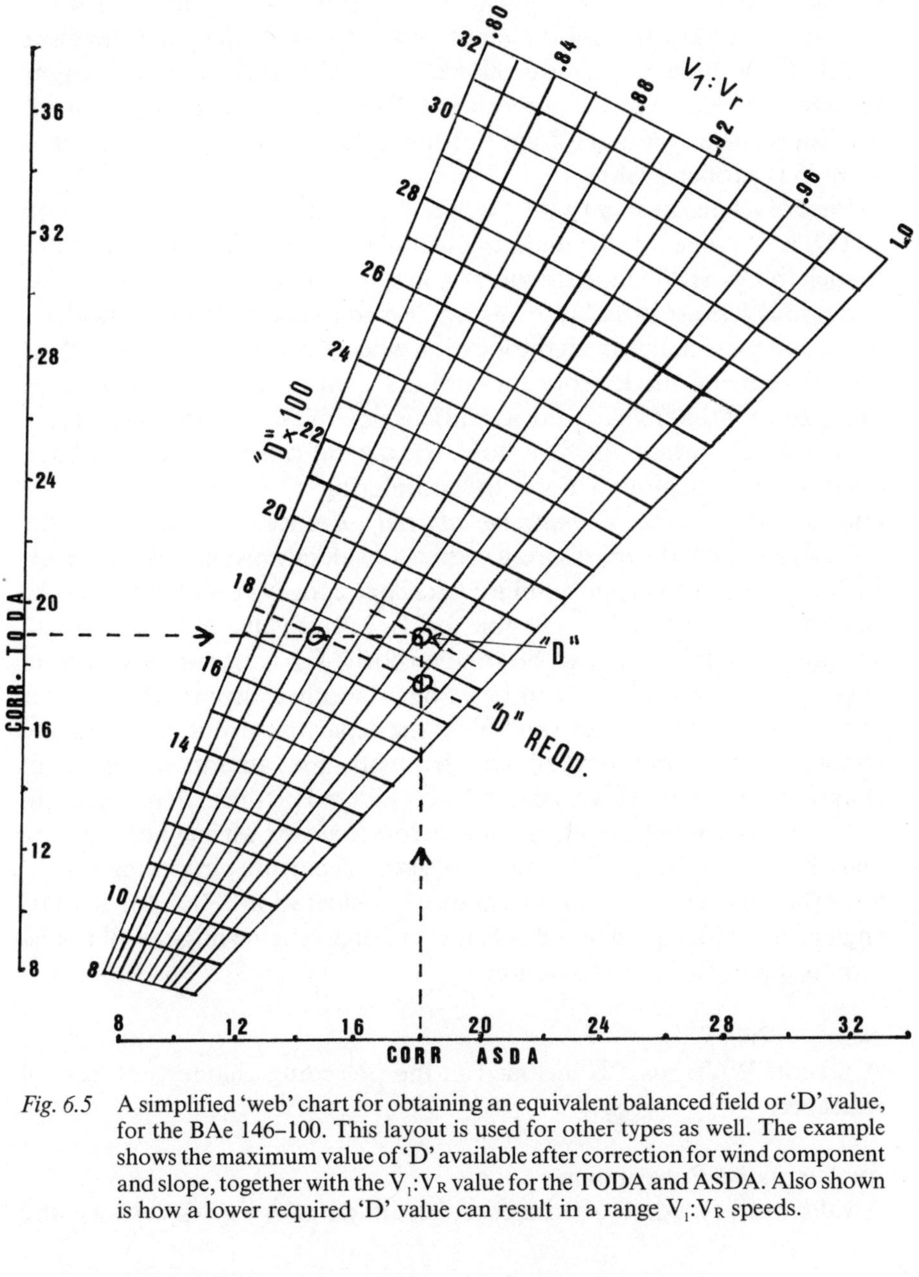

Fig. 6.5   A simplified 'web' chart for obtaining an equivalent balanced field or 'D' value, for the BAe 146–100. This layout is used for other types as well. The example shows the maximum value of 'D' available after correction for wind component and slope, together with the V_1:V_R value for the TODA and ASDA. Also shown is how a lower required 'D' value can result in a range V_1:V_R speeds.

For the TOW and ambient temperature at msl as quoted above, the V_r is 121 kt. V_1 may therefore range from 105.5 kt (0.87 ratio) to 117 kt (0.96 ratio). This means that the aircraft satisfies the climb-out with engine failure from a V_1 of 105.5 kt (while still satisfying the accelerate-stop requirement) *or* it may still continue its all-engines acceleration up to an engine failure speed (V_{EF}) that is appropriate to a V_1 of 117 kt. In other words, it reaches a V_{GO} situation at 105.5 kt and a 'V_{STOP}' situation at 117 kt. Above 117 kt the takeoff cannot be abandoned safely and therefore should be continued. The acceptance of the V_{GO} and V_{STOP} philosophy has been advocated by the author for well over the last decade, and it has now found its way into the Joint European Airworthiness Requirements, JAR 25 (October 1988).

First of all, the average pilot's viewpoint. A high-speed rejected takeoff (RTO) is often held to be undesirable and unnecessarily hazardous, due to such things as directional control, and tyre-bursting. A low-speed RTO is certainly less stressful but there are also sound commercial considerations, and these derive from the location where the RTO is involved. If the takeoff is being made from the airline's own base it is commercially preferable for the engine to fail as late as is possible during the takeoff run, for the simple reason that if it does, the takeoff can be aborted at base, close to rectification action. In considering this point it must be understood that the V_1 *must be adhered to*, once decided before the takeoff, and only the most serious other considerations can excuse a post-V_1 RTO. But, if the engine failure takes place away from base it can be held to be commercially preferable for the lowest value of V_1 to be used, and the aircraft allowed to become airborne (as is permitted) with an engine out. It may then fly to base, or to another airport where better service facilities than those available at the departure airport exist. But the sector distance and terrain, coupled with the aircraft's engine-out characteristics, must be very carefully considered before arriving at a firm policy. In the case of a twin, en route across the Alps, an engine failure on takeoff might indicate a landing back at the departure airport anyway, if airborne, because of terrain-clearance considerations. But for a four-engined aircraft, e.g. the BAe 146, the loss of an engine might well not be serious under such circumstances.

Note
A takeoff WAT chart is included in the preceding chapter, for ease of reference.

An Alternative Presentation
An alternative method of presenting takeoff data for a given runway, and

Table 6.3

DAN-AIR SERVICES LTD

TAKE-OFF PERFORMANCE
BAE 146 100 SERIES WITH ALF 502–R3 ENGINES

AIRPORT CODE:BRN;

BERNE/LSZB

BERNE
ELEV: 1674 FT
FLAPS 30

FLAP 30

RUNWAY		14	32
SLOPE	%	0.14	–0.15
TORA	M	1310	1310
ASDA	M	1310	1310
TODA	M	1510	1370

OBSTACLE   5

TEMP (C)	N1REF (%)	A & C		B & C		CLIMB LIMIT
NOTES						
		RUNWAY PERMITTED TOW & V1 (X 10 KG & KT IAS)				TOW (KG)
–6	90.6	3800	98	3800	101	3800
–4	91.0	3800	98	3800	101	3800
–2	91.3	3800	98	3800	99	3800
0	91.6	3800	98	3800	99	3800
2	92.0	3800	98	3796	100	3800
4	92.4	3800	98	3790	99	3800
6	92.5	3800	98	3783	99	3800
8	92.5	3800	98	3772	99	3800
10	92.5	3775	97	3718	98	3800
12	92.3	3725	97	3662	98	3800
14	92.1	3678	96	3611	97	3800
16	91.9	3630	95	3560	96	3800
18	91.7	3583	95	3510	96	3800
20	91.5	3537	94	3460	95	3800
22	91.3	3490	93	3409	95	3800
24	91.0	3447	93	3358	94	3800
26	90.7	3403	92	3308	94	3800
28	90.3	3359	92	3258	94	3743
30	89.9	3315	91	3207	93	3682
32	89.4	3274	91	3158	93	3615
34	88.8	3233	90	3109	92	3549
36	88.3	3193	90	3058	92	3483
38	87.9	3152	89	3008	91	3417

WIND COMP. CORR.

Head (KG/KT:V1/10KT)	89	3	28	3
TAIL (KG/KT:V1/10KT)	–495	–10	–532	–11
FLAP RETRACTION HEIGHT AAL	400		400	

MAXIMUM CERT TOW   37308 KG
14   OBSTACLE(S)   CONSIDERED, NOT LIMITING
32   OBSTACLE   5
       DISTANCE   M   2735
       HEIGHT   FT   190
DATE 02/02/87   151   152

MAXIMUM LANDING WEIGHT 33271 KG

Notes:
(A) <u>EMERGENCY TURN:</u> R/W 14 – At end of runway turn <u>LEFT</u> to track 138°(M) until 2500 ft., then <u>LEFT</u> turn to 'MURI' NDB 312 kHz. From 'MURI' track 330°(M) to 'SHU' NDB 356.5 kHz, climbing to MSA in the SCHUPBERG Hold.
(B) <u>EMERGENCY TURN:</u> R/W 32 – At end of Runway, turn <u>LEFT</u> onto heading 290°(M) to intercept and track 315°(M) from 'BER' NDB 366.5 kHz. To intercept 350°(M) into 'SHU' NDB 356.5 kHz, climbing to MSA in the SCHUPBERG Hold.
(C) Take-off from a standing start using Rated N1 Power.

AMENDMENT NO. 121 – 20.3.87

equally acceptable (containing as it does more data) is where a page is allocated to each airport. Table 6.3 shows how Dan-Air Services, a large UK operator, presents takeoff data for a single airport on a single page. In this layout there is space for four runways per page; more than that would involve one or more extra pages. The table refers to a BAe 146-100, and the airport is Berne (Belp) in Switzerland. As in the previous presentation the ICAO location indicator is given (LSZB), and also the IATA code BRN. The chart is drawn up for a single flap setting, (30°), this being the optimum for the airport. It will be noted that the elevation is 1674 ft and that there are two runways, these being 14 and 32. The TORA and ASDA for R/W 14 are the same at 1310 m, with TODA at 1510 m. Slope is 0.14%. The minus sign preceding the slope value for R/W 32 shows that this is *up*, and that therefore the 0.14% value for R/W is *down* (the minus sign indicates that it has a reducing effect on 'D'). In the case of R/W 32 the TORA and ASDA are both 1310 m and there is 60 m clearway available providing a TODA of 1370 m.

It will next be noted that there are no obstacles that affect R/W 14 but that there is a 190 ft obstacle in the case of R/W 32. The obstacle data at the foot of the table shows that this is 2735 m along the takeoff path. Tabulated for each R/W are values of OAT, together with V_1 for each. An engine power reference for takeoff, N1%, is provided for each temperature. The extreme right-hand column shows the WAT limited weight for OAT; it will be noted that WAT does not limit until the value of OAT reaches +28°C. This WAT limited weight is 37 430 kg and this is above the *current* max. auw approved of – 37 308 kg. (The latter weight will be found below the table.) A wind component correction is given following the RTOW columns, which is valid for still air (W/C=0) only. In the case of R/W 14, each knot of headwind increases the RTOW by 89 kg, while the V_1 value increases by 3 kt per 10 kt headwind W/C. A tailwind reduces the RTOW by 495 kg per knot, and the V_1 by 10 kt per 10 kt tail W/C. The effects of W/C are more adverse in the case of R/W 32. Flap retraction height is 400 ft for the appropriate flap retraction speed V_{FR}. As previously mentioned, this method of presenting takeoff data enables much more useful information to be provided, at the expense of more pages. Probably a combination of the two methods is useful, in that the first allows for the crew to prepare their own data at a strange airport by reference to the 'web' chart alone in the AFM, so as to obtain 'D' values and the ratio $V_1:V_r$.

7: Flight Planning

We now reach a point in our study of the contents of a route book when we need to consider carefully what it is that we are about to examine, and to define this clearly. Flight planning in isolation is, in fact, a subject in its own right and has formed the basis for at least one textbook devoted solely to this matter. Flight planning can be used as another phrase for preflight navigational preparation and procedures. But we are going to look at an even earlier stage of this subject, namely the provision of data and information that will be used in the flight planning process defined above. And, in this context, flight planning forms the next chapter in the route book.

General

Most aircraft types – at least as regards transport-category aircraft, which are what concern us – will have varying performance characteristics in operation. If we go back to Fig. 2.1 we will see a typical flight profile, as practised in normal airline use. Every aircraft type will have different performance characteristics from takeoff to landing; some will be at their best using a steep gradient of climb and descent, with a long cruise element, while others are more suited to a more shallow profile, with a shorter cruise length. Yet others – especially the short-haul types – use a trajectory-type profile. That is to say, a rapid climb to high altitude that is followed after only a few minutes of cruise by a steep descent. Ambient temperature, en route winds and weight, all play their part in varying the route performance. In general, though, it may be said that both turboprops and turbojets need the higher altitudes, or flight levels, for their cruise. In fact, it is quite normal practice for the cruising flight level (FL) to be increased as weight reduces due to fuel burnoff, especially on the longer haul and less congested routes. The author remembers one of the earlier long-haul turboprops (the Britannia 310) where the temperature:weight relationship during a long flight indicated not infrequent changes in the optimum FL, either up or down, usually due to changes in the OAT and the need to conserve fuel. Such changes in FL caused little problem on the non-congested routes involved, but had they been requested from ATC on the more heavily used routes it is likely that some at least would have been denied. And what the airline wants is the minimum possible burnoff and the shortest time.

The aircraft manufacturer naturally tries to design the aircraft to meet the widest spectrum and there has, of necessity, to be some compromise. As some means of helping all operators of the aircraft type the manufacturer will provide very detailed operating data indeed, thereby enabling the operator to ascertain its own optimum flight procedures for each route. Such data can be so detailed as to be virtually unsuited for anything other than ground (route planning) use.

But it is always better to have too much data rather than too little, and the answer is for this to be reduced, in the course of a once and for all operation by route planning, into ready-use data. This should be in a format that can be rapidly used by either flight crew or ground flight despatch. Bearing in mind the very strictly enforced and stringent limits, on flight crew duty and flight times, it is quite unacceptable to involve them in long preflight procedures. And so simplified procedures and data must be made available to enable the required sector fuel and flight time to be arrived at with the minimum of effort and time. It is the purpose of this chapter to illustrate how to collate and present all the necessary data from which a flight may be planned and a flight plan prepared, accurately.

A lot of this function is now being increasingly entrusted to computers. But it is the opinion of the author that, while these instruments are now quite reliable, undetected errors *could* creep in. It happens often enough in other walks of life! The author personally recalls using a small, hand-held electronic navigation computer that persistently calculated a rhumb line track that was consistently wrong by several tens of degrees, but self-evidently so, mercifully. Therefore it is his view that it must remain essential for operating crew to not only have written data available but also to know how to use it. An illustration of non-computer based techniques is described in the 'Route Details' section of this chapter and a computerised plog is illustrated in Fig. 7.1.

Normally an aircraft manufacturer will provide either tabular or graphical data, covering the takeoff and climb, cruise, descent, and landing, for varying conditions and parameters. As mentioned, OAT and weight must be accounted for, as must be the flight needs. For example, does the nature of the route indicate the use of high speed (HSC) or long range cruise (LRC) procedure? And, is it the departure to destination stage, or a diversion to an alternate that is being considered? And what about the holding procedure? And so on and so forth. So, one of the first functions of route planning in this context is to establish optimum flight procedures, probably in co-operation with the manufacturer, and then to prepare simplified data, specifically for its own operational pattern, from that supplied by the manufacturer. This can be, and often is, a highly laborious task. But, once completed, it is a thing of the past, usually. But, a word of warning. Sometimes the manufacturer can be over-academic in his approach to the subject of on line operations. The author recalls one very highly respected aircraft manufacturer who provided a route analysis that was based on optimum altitudes, *irrespective of the rules of the air*! Quadrantal heights were ignored, and the cruising FL values were shown to the nearest ten feet! The cause of this total ignorance of the ICAO rules applicable to cruising flight levels was simply that the manufacturer's computer literally selected the true optimum FL, having been programmed thus, as opposed to the nearest permissible, under international rules of the air.

A flight level is an altitude *indicated* by the aircraft's altimeter when the subscale is set to 1013.2 millibars, or 29.92 in. Hg.

Flight levels for cruising are allocated by air traffic control and all aircraft under ATC should have their altimeters set to the same pressure standard. Thus, a flight level is not necessarily an altitude above msl or above the ground, but is a standard, the base of which is *normally* well clear of the ground, except in mountainous regions. Cruising FLs are allocated on a directional basis, i.e. easterly tracks (from 0° to 179°) are allocated odd values while westerly tracks (180° to 359°) are given even values. This practice means that traffic on conflicting headings should always have at least 1000 ft clearance between such aircraft – e.g. an aircraft heading, say, 97° to make good a track 100° may be flying at FL 210, while a second aircraft on a reciprocal track may be at either FL 200 or FL 220. Up to FL 280 and FL 290, according to the track, cruising FLs are allocated at 2000 ft intervals in each 180° arc defined above; above these FLs (known as low level) the intervals are doubled to 4000 ft. Thus, in the easterly segment the quadrantal FLs are, say, FL 70, 90, 110, 130, etc. while the westerly FLs will be FL 60, 80, 100, 120 and so on, up to FL 290 or FL 280, as appropriate to the track. Above the two latter FLs, the quadrantal separation increases from 2000 ft to 4000 ft, giving upper level (or high altitude) FLs such as FL 280, 320, 360, and so on. (A useful mnemonic for remembering the rule is *'E for East, E for Evens – that's Odd!'*)

Every aircraft type, and even individual aircraft within small limits, has its optimum cruise FL, according to OAT and weight. Normally, as the weight reduces as a function of fuel burnoff, the optimum FL will increase. However, the weight will need to reduce sufficiently to make an increase in optimum FL by at least 2000 ft valid, before any possible use can be made of this. And, if the original cruising FL was above FL 280 then the instantaneous weight must permit an increase to 4000 ft higher. All increases (or reductions) in FL must be approved by ATC, of course.

There is another factor, other than weight, that can also affect a change in optimum FL, and this is OAT. If flying into a warmer air mass it is quite possible that the optimum FL will not increase with burnoff, the OAT increase cancelling out the weight reduction effect. All of this depends upon the aircraft's characteristics, and it may even be the case that a *reduction* in optimum FL is indicated should the OAT increase be sufficiently great. But most modern airliners are usually subject to the weight rule.

Route Details

Before continuing with the subject of operating data let us turn our attention to the airline's network of routes, initially expressed as being between place names, e.g. London (Gatwick) to Berne. These will most certainly not be planned as straight lines (although, with modern ATC developments, it is possible that it could be on occasion – at least in part, anyway). The standard route will, in fact, be planned according to ATC requirements as normally applicable, and will probably commence with a standard instrument department (SID), joining airways at a convenient or designated, navigational facility. It will then follow the shortest applicable airways and/or advisory route (ADR) and may even include some element of off airways and ADR routing. At the end it will include a standard terminal approach route (STAR) and instrument approach procedure (IAP). The whole route will be prescribed in the route book in detail, normally in the form of a number of 'legs' between radio navigational facilities. The layout will, in essence, comprise a series of 'from' and 'to' legs, with the frequency and identification of the aid being aimed for given, together with its geographical co-ordinates and possibly a number (for area navigational purposes). Also given will be the minimum safe altitude (MSA) for the leg, and the track and distance together with a standard time. The latter will have been calculated using statistical winds, and the W/C used for the overall route will be clearly shown.

Sometimes the 'aiming point' referred to above will not be marked by any form of on-track navaid but will be identified by means of VOR radials, or a VOR radial with a distance measured by a distance measuring equipment (DME) co-located with the VOR. Or it may be that two airborne direction finders (ADF) will take bearings on two non-directional beacons (NDB). In all of these cases the point being sought will be a reporting point (RP) having no on-track locating aid. It may be assumed that all the points listed in the route details will be RPs, i.e. the aircraft will (probably) be required to report to ATC when overhead the RP. As mentioned above, route details will, or may, be contained in the route book. However, it may well be acceptable for these details to be printed in the form of a prepared log (often called a 'plog'), along with other information, such as communications frequencies. The plog must then be considered to be a part of the operations manual, even if the route data is also contained in the route book volume. Space is contained in the plog for recording revised times and required fuel load, due to non-standard forecast conditions, and also minimum sector fuel (MSF). Even if the required fuel load is less than the MSF at least MSF *must* be loaded preflight. Provision is also made on the plog for recording actual times, such as actual time of departure (ATD), and actual time over (ATO) – this being the actual time in GMT overhead the RP or waypoint. The forecast time for each leg should be recorded, alongside the standard time that is printed on the plog, while provision is made for recording fuel burned on each leg together with an estimate of the fuel remaining at the estimated time of arrival (ETA). The latter must always be at least equal to the minimum fuel required for reserve. The plog will have provision for showing how the fuel loaded is allocated, and a fuel log for showing actual consumption versus planned.

Flight Planning Methodology
At this point it would perhaps be wise to briefly discuss the initial methodology used in the preflight process. This must inevitably be associated with the type of route being flown, namely short, medium or long haul. Basically there is one overriding characteristic: the need to adhere as closely as possible to the optimum flight profile for the aircraft type being considered. All types of routes are subject to the effects of meteorology, and also to ATC considerations. The prime differences between the route types is that, in the main, the short and medium haul routes are operated using point-source navigation aids, such as VOR or NDB, although there is an increasing trend towards backing this method up by using an area-navaid such as Omega. Although the flight plan and standard route may be presented using airways, ADRs or direct tracks in between navaids, very often, especially over Western Europe, a flight may report overhead an aid/RP early en route and then receive a reclearance to a distant aid/RP, by-passing several waypoints (i.e. navaids and RPs), using a straight line between the RP from which the reclearance is issued to the next designated waypoint under this new cleared route, which may be several hundreds of miles ahead. For example, the author was recently flying as supernumerary crew on an Austrian Airlines flight between Salzburg and London. The flight plan had been filed using the normal airways routing. However, on reporting over Allersberg (near Nürnberg), ATC issued a 'direct-to' clearance to Koksy (near Ostend, on the Belgian

coast). This would normally entail drawing in the direct track and then measuring the angle on the navigation chart. With area navigation a long direct track and track keeping, requires very little, if any, calculation. With the aircraft in question having Omega coupled to a flight management system (FMS) and also to the autopilot and auto-throttle, the only action necessary on receipt of the reclearance was to key in the new waypoint no. (referred to earlier) on the FMS, press the execute button, and the aircraft automatically turned on to the required heading and also maintained the required track. In addition, because of the operator's policy, the new routing meant a shorter flight time and this saving was automatically addressed to the auto-throttle system, reducing thrust and therefore fuel, while maintaining the schedule.

The *routing* case of the long haul aircraft is somewhat different, but again this will depend very largely on the route being flown. For example, if we take the route London to Sydney, Australia, there will be at least one en route stop, both for traffic and refuelling purposes. The two separate legs will, almost certainly, comprise operations using a considerable amount of airways or ADR flying. But, if we take the London to New York route almost the converse will apply: the flight will be on airways to the west of Scotland or Ireland using ground, point-source navaids, and then the Atlantic crossing will be made using the on-board area navigation equipment although, admittedly, area navigation may rely on ground-based transmissions. On approaching the North American seaboard the flight will return to airways for the final stage of the flight. The actual route to be flown on the long Atlantic haul varies according to the density of traffic (ATC requirements and separation standards), and the pressure pattern of the air mass. ATC will have issued the prescribed routing, track, and FLs pre-flight, and the flight plan will then have to be prepared. This is now normally done by computer, using the forecast upper winds and OATs, but it might be that, on occasion, the flight planning will be carried out by human hand. This basically involves applying the aircraft's en route performance data to the route clearance so that it can be navigated to arrive at nominated geographical locations at prescribed heights (FLs), and within the precalculated fuel burnoff values for each point. Normally an area navigation system will be used to make good the required track. And, of course, as the aircraft becomes lighter, the commander will try to obtain reclearances to more optimum FLs.

Thus, apart from the various route distances involved, the basic principle of operation remains the same on short, medium, and long haul routes, namely, to aim for optimum altitude. However, on the first two types of route at least, there is not the same need to worry about the pressure pattern and en route winds. To simplify the flight planning process and prepared data, one method of taking the en route winds into account is to precalculate their effect, using statistical values. This is done in the following manner.

More and more airlines are changing over to computer-produced plogs. In these, an appropriately programmed computer will produce a plog for a given stage, using the applicable conditions as appropriate to

```
DAN    G-SCHH  LGW-BRN-1   EPN 60  APN____   DATE 26/01/89 11:46
CO FREQ 135.75(CROSSAIR)

EZFW 28417     ETOW 32860     ELW 30041      ALL WEIGHTS KGS

DIST  496NM  COMP   -1  CORR    AV TR 130
A-DEST       2819  1:26   _____                   DIST  TK  COMP TIME FUEL
B-CONT        141         _____   LFSB BASLE       82  005   -0   18  643
C-ALT         643   :18   _____   LSGG GENEVA      74  240    1   17  592
D-HOLD        840   :30   _____   LSZH ZURICH      79  045   -1   18  626 ⌐
E-TAXI/APU    200         _____
F-REQUIRED   4643  2:14   _____
                                  STA____  ON_____  LAND_____
        FOB _____
                                  STD____  OFF_____  A/B_____
CMR (C+D)    1483
MDF (C+J)    1063      CAPT_____  TOT_____  TOT_____
CLEARANCE                         ATIS
```

POS	WPT	LAT LONG	AWAY	MSA	FL	T(M)	SD	ST	ETA	RTA	ATA	REQ/FOB	FREQ
		FL COMP	IDENT FREQ										
EGKK	_	N51092 W000105											
WOR	_	N50421 W000150		SID	___	- - -	34	8	____	___	___		_____
		CLM 1											
VEULE	_	N49514 E000372	DCT	2.8	___	-152-	61	11	____	___	___		_____
		CLM 1											
T-O-C	_		UA1	2.8	___	-152-	20	3	____	___	___	32/___	_____
		CLM 1											
MAN	_	N48584 E001294	UA1	2.8	___	-152-	43	7	____	___	___	30/___	_____
		290 -2	MAN 112.0										
RBT	_	N48392 E001597	UA1	2.8	___	-138-	28	4	____	___	___	28/___	_____
		290 3	RBT 114.7D										
BRY	_	N48244 E003177	UG4	2.8	___	-109-	54	8	____	___	___	26/___	_____
		290 8	BRY 114.1D										
TRO	_	N48151 E003578	UG4	2.7	___	-114-	28	4	____	___	___	25/___	_____
		290 4	TRO 116.0										
RLP	_	N47544 E005150	UG4	7.5	___	-115-	56	9	____	___	___	22/___	_____
		290 4	RLP 117.3D										
LUL	_	N47413 E006178	UG4	17.8	___	-109-	44	7	____	___	___	20/___	_____
		290 -1	LUL 117.1										
T-O-D	_		G4	6.6	___	-116-	11	2	____	___	___	19/___	_____
		290 -1											
HR	_	N47337 E006441	G4	6.6	___	-116-	8	1	____	___	___		
		DSC -10	HR 289										
HOC	_	N47281 E007400	G4	7.2	___	-101-	38	6	____	___	___		_____
		DSC -10	HOC 113.2										
WIL	_	N47108 E007544	R73	7.2	___	-152-	20	4	____	___	___		_____
		DSC -10	WIL 116.9										
SHU	_	N47014 E007235	DCT	8.6	___	-248-	23	4	____	___	___		_____
		DSC -10	SHU 356.5										
SHU	_	N47014 E007235	IAC	___		- - -	20	4	____	___	___		_____
		DSC -10	SHU 356.5										
MUR	_	N46568 E007281	IAC	___		- - -	6	2	____	___	___		_____
		DSC -10	MUR 312										
LSZB	_	N46548 E007300	IAC	___		- - -	2	2	____	___	___	15/___	_____
		DSC -10											

Fig. 7.1   A typical computerised plog layout as used by Dan-Air Services.

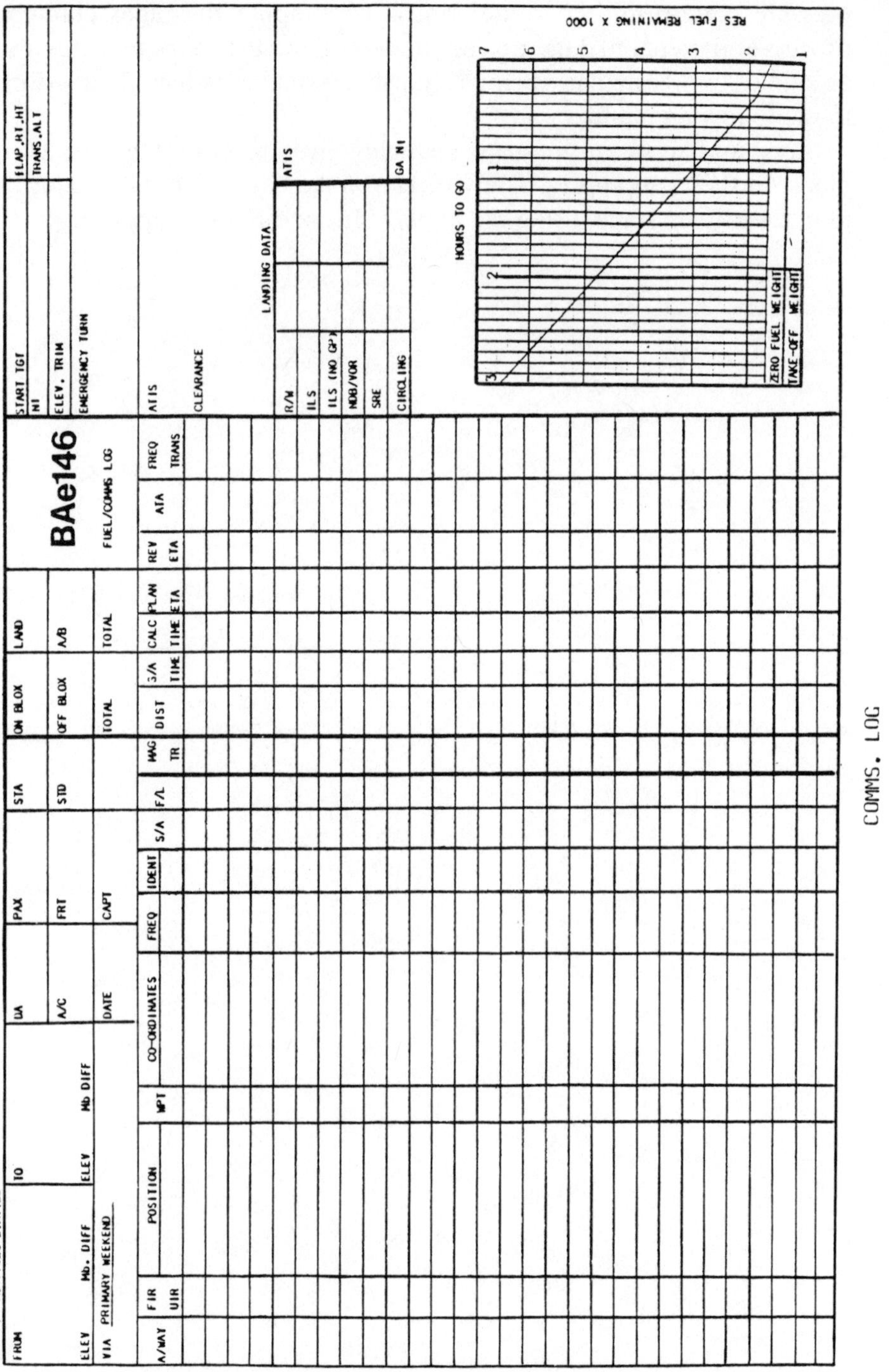

Fig. 7.2(a)   A manual plog as used by Dan-Air Services. (Courtesy of British Aerospace and Dan-Air Services Ltd.)

FUEL REQUIREMENTS BAe146

FROM

TO

ALTERNATE A/D HRS: H24 - 24 hrs   H24J - 24 hrs + Jet Ban
R = Restricted   P = Prior Permission

EN ROUTE WEATHER

ALTERNATES	AD HRS	AV TR	DIST	W/V	W/C	EET	FUEL

ROUTE	SA	AV TR	DIST	SERIES	FL	S/A FUEL	S/A TIME

PRIMARY WEEKEND
(SUMMER 86)

	FUEL	TIME
A DESTINATION		
B 5% × A or H		
C ALTERNATE (inc 5%)		
D HOLDING	840	00:30
E TAXY / APU	200	
F REQUIRED RAMP		
G EXTRA CONT.		
H TOTAL RAMP		
HBF (C + J)		
OHR (C + D)		

HOLDING

MINS	FUEL
15-J	420
30-D	840
60	1680

* CHECK A/D HRS

ROUTE WINDS	WIND COMP
	+
	-

EN ROUTE ALTERNATE

EN ROUTE ALTERNATE	
CHECK POINT	
DIST TO DEST.	
WIND COMP.	
FUEL TO DEST - M	

| TOTAL | |
| AVERAGE | |

STATION

COMMS. LOG (REVERSE)

Fig. 7.2(b)   The reverse side of Dan-Air's manual plog. (Courtesy of British Aerospace and Dan-Air Services Ltd.)

each flight. On a scheduled service, where a short turn-round is the norm, the computer will also supply a return plog, although this may need to be adjusted by the crew prior to the return takeoff should the conditions prevailing differ from those on the outward flight. And where the turn-round may involve a longer time, say, a night stop, the return plog will be prepared for a set of standard conditions, as close as is possible to those expected. Naturally, should the computer become unserviceable, a back-up system must be available, and this normally comprises a return to the older type of plog, and precalculated means of working out fuel requirements and time. This type of presentation has been referred to earlier.

In Fig. 7.1 a computer-produced plog is shown, this being for the Gatwick to Berne route, outbound. In this example the computer has calculated fuel requirements and times for a set of stated conditions. It will be noted that the W/C is –1, the distance is 496 nm, and the average Tr. (T) is 130°. The fuel breakdown is shown, giving a total sector fuel requirement of 4643 kg, which includes holding, diversion, taxying/APU and contingency. The total time allowed for is the sum of the stage time, diversion, and holding, amounting to 2 hours 14 minutes.

To cover those cases where a computer breakdown occurs, reversion must be made to 'old' methods of calculating fuel and time. Here the plog comprises a printed form, upon which the route details are inserted just as they are on the computed plog. Figure 7.2 (a and b) shows such a form of plog, but in blank. The procedure is then as follows.

Initially, the prescribed route details are entered on to the navigation log form. Then the Average True Track (Av. Tr. (T)) between the departure and destination is obtained (in Fig. 7.1, this is 130° (T)). This is done by drawing a straight line on the navigation chart between these two points and then measuring its angular relationship to true north at the midway meridian of longitude. The actual airways route may 'wriggle' about on either side of this line, but it denotes the *average Track* (Tr.). Next, the route statistical wind component is ascertained and from this the average head of tailwind component along the entire stage can be calculated by means of the statistical wind component. By using the W/C applicable to the Av. Tr. (T)), the standard time and burnoff can be obtained. (It is suggested that the 50% wind be used for time, and the 85% wind for fuel.) However, to carry out this process it is desirable that precalculated time and fuel sources be constructed. These can also be used for preflight planning and for the provision of commercial estimates.

To construct a typical time and fuel chart, suitable for short and medium stages, proceed as follows. First, pick a very short stage, say, 75–100 nm and ascertain the optimum altitude, or flight level (whichever is the more appropriate). Initially use zero wind and ISA and calculate the time, distance, and fuel burnoff for the takeoff and climb to cruise FL. Repeat this process for W/C values, at 10 knot intervals, for both head and tail components, up to 60 kt or so. The W/C value used for each climb should be modified before calculating so as to use only two-thirds of the nominal component. For example, in constructing the 30 kt climb, use 20 kt W/C. The result at the end will be a family of values all valid for a climb to the same altitude but differing as regards distance, as a result of the W/C used. Naturally, it will take the same time, and use the same amount of fuel to reach a given FL.

Having thus calculated climb times, fuel and distances, now repeat the process for the descent, but only using 50% of the nominal W/C value, say, 15 kt for a 30 kt W/C. Once again, a family of descent profiles will result, with only the distance covered differing for each. Take each climb and descent calculation for a notional W/C value, and add these in turn – that is, add the 10 kt headwind climb and descent, the 20 kt climb and descent, and so on. Now subtract the *distances* so totalled from the total stage length selected, and this will give the cruise distance using the full W/C value. (Incidentally, the weight assumed for the takeoff should be based on maximum landing weight (MLW) plus the *estimated* burnoff, with a standard average reserve, and an average estimated cruise weight, namely estimated mid-cruise weight.) The cruise data will comprise a TAS value and a rate of fuel flow. Knowing the cruise distance, the TAS and the flow, we can calculate the time of cruise, according to the TAS, plus or minus the W/C, and thus the fuel burned for that time from the total flow per hour.

Take as an example the BAe 146-100 flying a stage distance of 500 nm; the temperature is ISA. The track is westerly, so that a flight level for the cruise element will be an even value (say, FL 280). Assume for this example a 0 W/C for simplicity. The max. landing weight is 35 153 kg and the *estimated* burnoff for the stage is 2500 kg. This gives an *estimated* TOW of 37 653 kg, and a mid-cruise weight of 36 403 kg. Now turn to Table 7.1 and obtain the climb values for the TOW 37 653 kg to FL 280, interpolating between the weights 37 000 kg and 38 000 kg. It will be seen that the time is only 1 minute more, while the distance covered between the two weights is 6 nm more for the higher, while 50 kg more fuel are burned. We can thus allocate ⅔ (roughly) more fuel to the 37 000 kg TOW value by adding 4 nm and 34 kg, thereby obtaining 1124 kg fuel, 98 nm distance, and 20 mins for time, all for the climb element. Now repeat for the descent, estimating the weight for the top of the descent (TOD) from the burnoff estimated at 2500 kg, less the fuel required for the descent, using Table 7.2. This is far less critical than the climb and we can take the

Table 7.1 Flight planning BAe 146 operations manual. Climb – long range: ISA. (Courtesy of British Aerospace.)

WEIGHT AT START OF TAKE-OFF (KG)

PRESS ALT FEET	28000 FUEL KG	DIST NM	TIME MIN	30000 FUEL KG	DIST NM	TIME MIN	32000 FUEL KG	DIST NM	TIME MIN	33000 FUEL KG	DIST NM	TIME MIN	34000 FUEL KG	DIST NM	TIME MIN	35000 FUEL KG	DIST NM	TIME MIN	36000 FUEL KG	DIST NM	TIME MIN	37000 FUEL KG	DIST NM	TIME MIN	38000 FUEL KG	DIST NM	TIME MIN	PRESS ALT FEET
31000	780	71	14	860	79	16	950	89	18	1000	94	19	1050	99	20	1110	106	21	1180	113	22	1250	120	24	1330	129	25	31000
30000	750	67	14	830	75	15	920	83	17	960	88	18	1010	93	19	1070	98	20	1120	104	21	1190	111	22	1260	118	23	30000
29000	730	63	13	800	70	14	880	78	16	930	82	17	970	87	18	1020	92	19	1080	97	20	1130	102	21	1200	109	22	29000
28000	700	60	12	780	66	14	850	73	15	890	77	16	940	81	17	980	85	18	1030	90	19	1090	95	19	1140	101	20	28000
27000	680	56	12	750	62	13	820	69	15	860	72	15	900	76	16	950	80	17	990	84	18	1040	88	18	1090	93	19	27000
26000	640	53	11	720	58	12	790	64	14	830	67	14	870	71	15	910	74	16	950	78	16	1000	82	17	1050	87	18	26000
25000	630	49	11	700	54	12	760	60	13	800	63	13	830	66	14	870	69	15	910	73	15	960	77	16	1000	80	17	25000
24000	610	46	10	670	50	11	730	55	12	760	58	13	800	61	13	830	64	14	870	67	15	910	70	15	950	74	16	24000
23000	580	42	9	640	46	10	690	51	11	720	53	12	760	56	12	790	59	13	830	61	14	870	64	14	910	68	15	23000
22000	550	39	9	610	43	10	660	47	11	690	49	11	720	51	12	750	54	13	790	56	13	820	59	13	860	62	14	22000
21000	530	36	8	580	39	9	630	43	10	660	45	10	690	47	11	720	49	11	750	52	12	780	54	12	820	57	13	21000
20000	500	33	8	550	36	9	600	40	9	630	42	10	650	43	10	680	45	11	710	47	11	740	50	12	770	52	12	20000
19000	480	30	7	520	33	8	570	37	9	590	38	9	620	40	10	650	42	10	680	44	11	700	46	11	730	48	11	19000
18000	460	28	7	500	31	8	540	34	8	560	35	9	580	37	9	610	38	9	640	40	10	660	42	10	700	43	11	18000
17000	430	26	6	470	28	7	510	31	8	540	32	8	550	33	8	580	35	9	610	37	9	630	38	10	660	40	11	17000
16000	410	23	6	450	26	7	490	28	7	510	29	8	530	31	8	550	32	8	570	33	8	600	35	9	620	36	10	16000
15000	390	21	6	420	23	6	460	26	7	480	27	7	500	28	7	520	29	7	540	30	8	570	32	8	590	33	9	15000
14000	370	19	5	400	21	6	430	23	6	450	24	7	470	25	7	490	26	7	510	27	7	530	29	8	550	30	8	14000
13000	350	16	5	380	19	5	410	21	6	430	22	6	430	23	6	460	24	6	480	25	7	500	26	7	520	27	7	13000
12000	320	16	5	350	17	5	380	19	5	400	20	6	410	20	6	430	21	6	450	22	6	470	23	7	490	24	7	12000
11000	300	14	4	330	15	5	360	17	5	370	18	5	390	18	5	400	19	5	420	20	6	440	21	6	460	21	6	11000
10000	280	13	4	310	14	4	330	15	5	340	16	5	360	16	5	380	17	5	390	18	5	410	18	6	430	19	6	10000
9000	240	11	4	290	12	4	310	13	4	320	14	4	330	14	4	350	15	5	360	15	5	380	16	5	390	17	5	9000
8000	240	10	3	260	10	3	290	11	4	300	12	4	310	12	4	320	13	4	340	13	4	350	14	5	360	14	5	8000
7000	220	8	3	240	9	3	240	10	3	270	10	3	280	11	3	300	11	4	310	11	4	320	12	4	350	12	4	7000
6000	200	7	3	220	7	3	240	8	3	250	8	3	260	9	3	270	9	3	280	10	3	290	10	4	300	10	4	6000
5000	180	6	2	200	6	2	220	7	2	220	7	2	230	7	3	240	8	3	250	8	3	260	8	3	270	9	4	5000
4000	150	4	2	170	4	2	180	5	2	190	5	2	200	5	2	200	5	2	210	5	2	220	6	2	230	6	3	4000
3000	130	2	1	140	3	1	150	3	1	160	3	1	160	3	1	170	3	2	180	3	2	180	4	2	190	4	2	3000
2000	100	1	1	110	1	1	120	1	1	120	1	1	130	1	1	130	1	1	140	2	1	140	2	1	150	2	2	2000
1000	70	0	–	80	0	–	80	0	–	90	0	–	90	0	–	90	0	–	100	0	–	100	0	–	110	0	1	1000

Table 7.2 Flight planning BAe 146 operations manual. Descent – long range. (Courtesy of British Aerospace.)

LONG RANGE DESCENT PERFORMANCE

ISA

ARRIVAL WEIGHT (KG)

HT FEET	38000 FUEL KG	DIST NM	TIME MIN	36000 FUEL KG	DIST NM	TIME MIN	34000 FUEL KG	DIST NM	TIME MIN	32000 FUEL KG	DIST NM	TIME MIN	30000 FUEL KG	DIST NM	TIME MIN	28000 FUEL KG	DIST NM	TIME MIN	HT FEET
31000	340	106	21	330	105	21	330	104	21	320	103	21	320	101	20	310	99	20	31000
30000	290	96	19	290	96	19	290	95	19	290	93	19	280	91	19	280	89	18	30000
29000	290	93	19	290	92	19	280	91	19	280	90	18	280	88	18	270	86	18	29000
28000	280	90	18	280	89	18	280	88	18	270	87	18	270	85	18	260	83	17	28000
27000	270	87	18	270	86	18	270	85	18	270	84	17	260	82	17	260	80	17	27000
26000	270	84	17	270	83	17	260	82	17	260	81	17	250	79	16	250	77	16	26000
25000	260	81	17	260	80	17	260	79	17	250	78	16	250	76	16	240	74	16	25000
24000	250	78	16	250	77	16	250	76	16	250	75	16	240	73	15	240	71	15	24000
23000	250	74	16	250	73	16	240	73	16	240	72	15	240	70	15	230	68	15	23000
22000	240	71	15	240	70	15	230	69	15	230	68	15	230	67	14	220	65	14	22000
21000	230	67	14	230	67	14	230	66	14	220	65	14	220	63	14	220	62	13	21000
20000	220	64	14	220	63	14	220	63	14	220	62	13	210	60	13	210	59	13	20000
19000	210	60	13	210	60	13	210	59	13	210	58	13	210	57	13	200	56	12	19000
18000	200	57	13	210	57	13	200	56	12	200	55	12	200	54	12	190	53	12	18000
17000	200	54	12	200	53	12	200	53	12	190	52	12	190	51	12	190	50	11	17000
16000	190	50	11	190	50	11	190	50	11	190	49	11	180	48	11	180	47	11	16000
15000	180	47	11	180	47	11	180	46	11	180	46	11	170	45	10	170	44	10	15000
14000	170	44	10	170	44	10	170	43	10	170	43	10	160	42	10	160	41	10	14000
13000	160	41	10	160	41	10	160	40	9	160	40	9	160	39	9	160	38	9	13000
12000	150	38	9	150	37	9	150	37	9	150	37	9	150	36	9	150	35	8	12000
11000	140	34	8	140	34	8	140	34	8	140	34	8	140	33	8	140	33	8	11000
10000	130	31	8	130	31	8	130	31	8	130	31	8	130	30	7	130	30	7	10000
9000	110	28	7	120	28	7	120	28	7	120	28	7	120	28	7	120	27	7	9000
8000	110	25	6	110	25	6	110	25	6	110	25	6	110	25	6	110	24	6	8000
7000	100	22	6	100	22	6	100	22	6	100	22	6	100	22	6	100	22	6	7000
6000	90	19	5	90	19	5	90	19	5	90	19	5	90	19	5	90	19	5	6000
5000	80	16	4	80	16	4	80	16	4	80	16	4	80	16	4	80	16	4	5000
4000	60	12	3	60	12	3	60	12	3	70	12	3	70	12	3	70	12	3	4000
3000	40	7	2	40	7	2	40	7	2	40	8	2	40	8	2	40	8	2	3000
1500	0	0	0	0	0	0	0	0	0	0	0	0	0	0	0	0	0	0	1500

Table 7.3 Cruise control BAe 146 operations manual. Cruise – long range. (Courtesy of British Aerospace.)

28 000 ft
Table 4G

CRUISING WEIGHT (KG)	TEMP. REL. TO ISA (DEG C) AMBIENT TEMP. (DEG C)		-20 -60	-15 -55	-10 -50	-5 -45	0 -40	5 -35	10 -30	15 -25	20 -20
	T.A.S. (KT)		344	348	352	356	360	363	367	371	375
38000	FUEL FLOW/ENGINE (KG/HR) N1 (%)		440 81.5	450 82.5	460 83.6	470 84.6	480 85.6	480 86.6	490 87.5	500 88.4	510 89.3
36000	FUEL FLOW/ENGINE (KG/HR) N1 (%)		430 80.5	430 81.5	440 82.5	450 83.6	460 84.6	470 85.6	470 86.5	480 87.4	490 88.3
34000	FUEL FLOW/ENGINE (KG/HR) N1 (%)		410 79.6	420 80.6	430 81.6	430 82.6	440 83.7	450 84.6	460 85.6	460 86.5	470 87.4
32000	FUEL FLOW/ENGINE (KG/HR) N1 (%)		400 78.7	410 79.7	410 80.7	420 81.7	430 82.8	430 83.8	440 84.7	450 85.6	460 86.5
30000	FUEL FLOW/ENGINE (KG/HR) N1 (%)		390 77.9	390 78.9	400 79.8	410 80.9	410 81.9	420 82.9	430 83.9	430 84.8	440 85.6
28000	FUEL FLOW/ENGINE (KG/HR) N1 (%)		370 77.1	380 78.1	390 79.1	390 80.0	400 81.0	410 82.0	410 83.0	420 84.0	430 84.8
26000	FUEL FLOW/ENGINE (KG/HR) N1 (%)		360 76.4	370 77.4	380 78.3	380 79.3	390 80.3	400 81.3	400 82.3	410 83.2	410 84.1

29 000 ft
Table 4H

CRUISING WEIGHT (KG)	TEMP. REL. TO ISA (DEG C) AMBIENT TEMP. (DEG C)		-20 -62	-15 -57	-10 -52	-5 -47	0 -42	5 -37	10 -32	15 -27	20 -22
38000	T.A.S. (KT) FUEL FLOW/ENGINE (KG/HR) I.A.S. (KT) N1 (%)		349 450 235 82.5	353 460 235 83.6	358 470 235 84.7	362 470 235 85.7	366 480 235 86.7	369 490 235 87.6	373 500 235 88.6	377 510 235 89.5	
36000	T.A.S. (KT) FUEL FLOW/ENGINE (KG/HR) I.A.S. (KT) N1 (%)		349 430 235 81.4	353 440 235 82.5	358 450 235 83.6	362 450 235 84.6	366 460 235 85.6	369 470 235 86.6	373 480 235 87.5	377 490 235 88.4	381 490 235 89.3
34000	T.A.S. (KT) FUEL FLOW/ENGINE (KG/HR) I.A.S. (KT) N1 (%)		349 420 235 80.5	353 420 235 81.5	358 430 235 82.6	362 440 235 83.6	366 440 235 84.7	369 450 235 85.6	373 460 235 86.5	377 470 235 87.4	381 470 235 88.3
32000	T.A.S. (KT) FUEL FLOW/ENGINE (KG/HR) I.A.S. (KT) N1 (%)		349 400 235 79.6	353 410 235 80.6	358 410 235 81.7	362 420 235 82.7	366 430 235 83.8	369 440 235 84.7	373 440 235 85.7	377 450 235 86.5	381 460 235 87.5
30000	T.A.S. (KT) FUEL FLOW/ENGINE (KG/HR) I.A.S. (KT) N1 (%)		349 390 235 78.8	353 390 235 79.8	358 400 235 80.8	362 410 235 81.8	366 410 235 82.9	369 420 235 83.9	373 430 235 84.8	377 440 235 85.7	381 440 235 86.6
28000	T.A.S. (KT) FUEL FLOW/ENGINE (KG/HR) I.A.S. (KT) N1 (%)		349 380 235 78.0	353 380 235 79.0	358 390 235 80.0	362 400 235 81.0	366 400 235 82.0	369 410 235 83.1	373 420 235 84.0	377 420 235 84.9	381 430 235 85.8
26000	T.A.S. (KT) FUEL FLOW/ENGINE (KG/HR) I.A.S. (KT) N1 (%)		349 370 235 77.3	353 370 235 78.3	358 380 235 79.2	362 380 235 80.2	366 390 235 81.2	369 400 235 82.2	373 400 235 83.3	377 410 235 84.2	381 420 235 85.1

(kg)

estimated fuel figure as being 294 kg. Deduct this from the stage burnoff estimate 2500 kg thus obtaining the figure 2206 kg. Deducting this latter weight from the TOW estimate 37 653 we obtain a t.o.d. estimated weight of 35 447 kg. Interpolating between 36 000 kg and 34 000 kg we obtain the following descent figures: fuel 280 kg, distance 88.5 nm, and time 18 mins. We now total the figures for both climb and descent, thus obtained, and arrive at 1404 kg fuel, 186.5 nm distance, and 38 mins time.

Having obtained the climb and descent totals for fuel, distance, and time, in still air we now need to find the values for the cruise element. This will, of course, be the total stage distance less the climb and descent totals. From Table 7.3 we need to extract a flow rate per hour and a true air speed (TAS). At FL 280 we find that the approximate mid-cruise weight is 36 500 kg, using the TOW and t.o.d. weights above, and interpolating. For 36 500 kg MCW, in ISA, the fuel flow is 465 kg per engine per hour, totalling 1860 kg/hr, with a TAS of 360 kt. The cruise distance is 500–186.5 nm in still air, or 313.5 nm. At 360 kt the time for this distance is 52 mins. At an hourly flow of 1860 kg, 1620 kg are burned in 52 mins. We can thus total the fuel and time figures obtained above to give the total takeoff to landing burnoff and time, thus:

Climb and descent	=	1404 kg fuel, 38 mins time
Cruise	=	1620 kg fuel, 52 mins time
Totals for Stage	=	3024 kg fuel, 90 mins time + 5% 151 kg = 3175 kg

This is, of course, the burnoff *including* the 5% contingency allowance over the 500 nm stage in still air. But we need to know now what the burnoff and time will be over the same distance but with headwinds (usually denoted by the sign –) at 10 kt intervals, and tailwinds (+) at the same intervals up to, say, 60 kt. It must be remembered that there is no wind effect upon the climb and descent times and fuel figures. The only thing that is affected is the distance covered; a head wind decreases the climb and descent distance and increases the cruise distance, while a tail wind has the opposite effect. Also remember that without precalculated data (or computers), the above process would need to be carried out for each and every flight.

So, let us now evolve a simple way of presenting the total stage fuel and time values for a 500 nm stage length. As we are now introducing wind effect we need to obtain a ground speed for the climb and for the descent (still air time v. dist.) but retaining the time to height and fuel burnoff figures. We also need the ground speed for time of cruise, derived from the TAS and W/C, and the flow. Remember that we are working on a 10 kt W/C separation interval, and that 66% (²⁄₃) of this only, is used for the climb and 50% ($\frac{1}{2}$) for the descent. The full value is used for the cruise, which merely entails adding or subtracting, the total value of the W/C from the cruise TAS to give ground speed, and thereby time and fuel-burn from flow. It is relatively simple to construct a table that will give the total burnoff and time from 60 kt tailwinds and 60 kt headwinds. Remember that in the climb and descent elements, we have the basic values time to and from FL, distance covered, and fuel burned. Irrespective of the W/C the climb and descent will always take the same times and burn the same amount of fuel. Only the distance is affected by the wind, and it is a very simple matter to obtain a still air TAS and G/S from the time and distance values for zero W/C. Having got this value (the two are, of course, the same) it is straightforward arithmetic to apply the factored (i.e. 66% or 50%) W/C to obtain a G/S appropriate to the mean W/C. We can therefore apply a G/S value to the constant time to obtain a

Table 7.4 Planning chart data

Aircraft: BAe 146–100   Temp.: ISA   Stage Distance: 500 nm

TOW: 37 653 kg   MLW: 35 153 kg   MCW: 36 400 kg   Cruise FL: 280   Av.CRZ TAS: 360 kt

Av. CRZ Flow: 1953 kt/hr*

Climb: S/A G/S 294 kt Fuel* 1180 kg   Time 20 mins
Descent: S/A G/S 294 kt Fuel* 294 kg   Time 18 mins

Full W/C kt	66% W/C kt	G/S kt	Climb Dist. nm	50% W/C kt	G/S kt	Descent Dist. nm	Total nm	CRZ Dist. nm	CRZ Time min	CRZ Fuel* kg	Stage Time min	Stage Fuel* kg
−60	−40	254	85.0	−30	264	79.0	164.0	336.0	67.0	2181	105.0	3655
−50	−33	261	87.0	−25	269	81.0	168.0	332.0	64.5	2099	102.5	3573
−40	−27	267	89.0	−20	274	82.0	171.0	329.0	62.0	2018	100.0	3492
−30	−20	274	91.5	−15	279	84.0	175.5	324.0	59.0	1920	97.0	3394
−20	−13	281	94.0	−10	284	85.0	179.0	321.0	55.5	1807	93.5	3281
−10	− 7	287	96.0	− 5	289	86.0	182.0	318.0	54.5	1774	92.5	3248
0	0	294	98.0	0	294	88.5	186.5	313.5	52.0	1701	90.0	3175
+10	+ 7	301	102.0	+ 5	299	90.0	190.0	310.0	50.0	1628	88.0	3102
+20	+13	307	102.5	+10	304	91.0	193.0	307.0	48.5	1579	86.5	3053
+30	+20	314	105.0	+15	309	93.0	198.0	302.0	46.5	1514	84.5	2988
+40	+27	321	107.0	+20	314	94.0	201.0	299.0	45.0	1457	83.0	2931
+50	+33	327	109.0	+25	319	96.5	205.0	295.0	43.0	1400	81.0	2874
+60	+40	334	111.5	+30	324	97.0	208.5	291.0	41.5	1351	79.5	2825

Notes
* All fuel figures *include* 5% contingency allowance, but no taxy allowance. Climb and descent distances are derived from still air time to FL and distance covered, giving a S/A G/S, which is then adjusted for the W/C (66% or 50%) to give appropriate G/S and thus distance for W/C. The *Total* climb and descent distances are then subtracted from the stage distance 500 nm, to give the cruise distance for the full W/C (left-hand column).

The climb W/C is 66% of the full W/C while 50% is used for the descent.

distance appropriate to the mean wind. Adding the climb and descent distances arrived at for the appropriate wind gives us a value to be subtracted from the nominal stage distance being used; in this case 500 nm. The distance remaining represents the cruise (CRZ) distance for each W/C, and the full 100%, component is used to obtain time, through G/S.

Table 7.4 illustrates how the calculation for a specific notional distance is carried out. The first column gives the full W/C value, with, following, the mean climb W/C (66%). Then follows the G/S (as appropriate to the climb TAS for still air, namely 294 kt. This is reduced or increased according to the factored value of W/C, e.g. for a 10 kt headwind the W/C used (66%) is about 7 kt headwind, giving a climb G/S of 287 kt. The constant climb time is 20 mins so that at 287 kt the distance covered will be 96 nm. The fuel value, being a function of time, also remains constant. The descent is dealt with in a similar way, except that the factored W/C is only 50% of the nominal value. The climb and descent distances are now added – remembering that the time and fuel values remain the same for all W/C values – and the totals are then noted in the form of increments. For convenience the 5% contingency allowance is included in the fuel figures; this also applies to the cruise (CRZ) values.

We now have climb and descent totals for each 10 kt deviation from zero wind component, up to +60 and down to –60. As time and fuel burn values are constant the only change due to the effect of the W/C is the value of distance that is covered during the climb and descent. We could, if we wished, drawn columns in which the values for time, fuel, and distance for climb and descent could be tabulated against each W/C value. But, if this were to be done each value of time and fuel would merely be repeated, being constants. It is simpler to note these values as two increments, or totals, to the main variables, i.e. time, fuel, and distance for the cruise element. The cruise distance is simply the notional stage distance less the distance covered, in total, for the climb and descent for each W/C value, remembering that the cruise W/C is the full value (100%).

We therefore deduct the climb and descent distances from the notional stage distance, e.g. 500 nm for the case under discussion, and note the resulting distance against the appropriate W/C. If we refer to Table 7.4, and take the first column (full W/C) for –60 kt we find that the climb distance is 85 nm, and the descent distance is 79 nm, making a total of 164 nm. By subtracting this total from 500 nm we find that the cruise distance is 336 nm. The average cruise TAS is 360 kt and the hourly fuel flow is 1953 kg/hr (all fuel figures in this Table include the 5% contingency allowance). With a headwind of 60 kt the time to fly 336 nm is 67.0

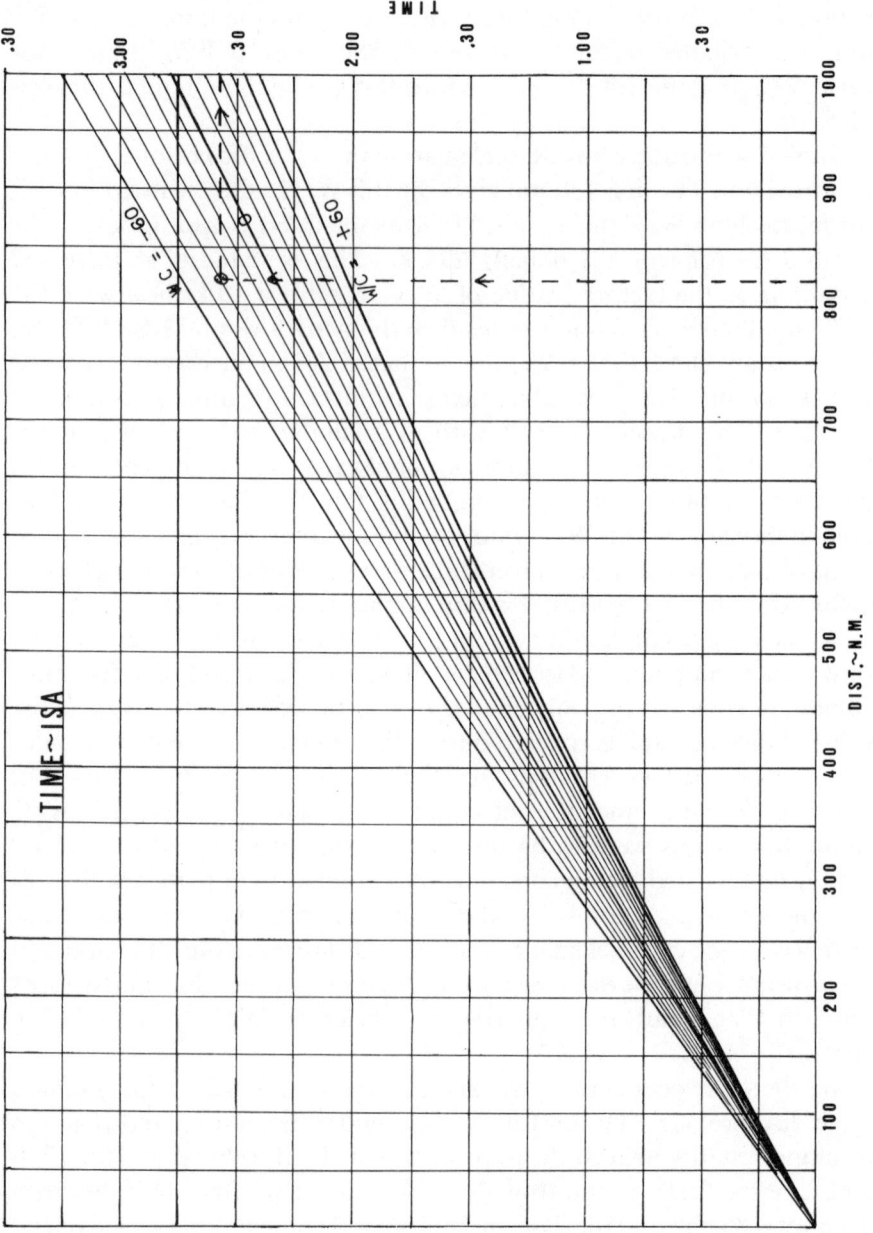

Fig. 7.3 One form of quick planning chart, providing takeoff to landing time for a modern airliner. No ground allowances included.

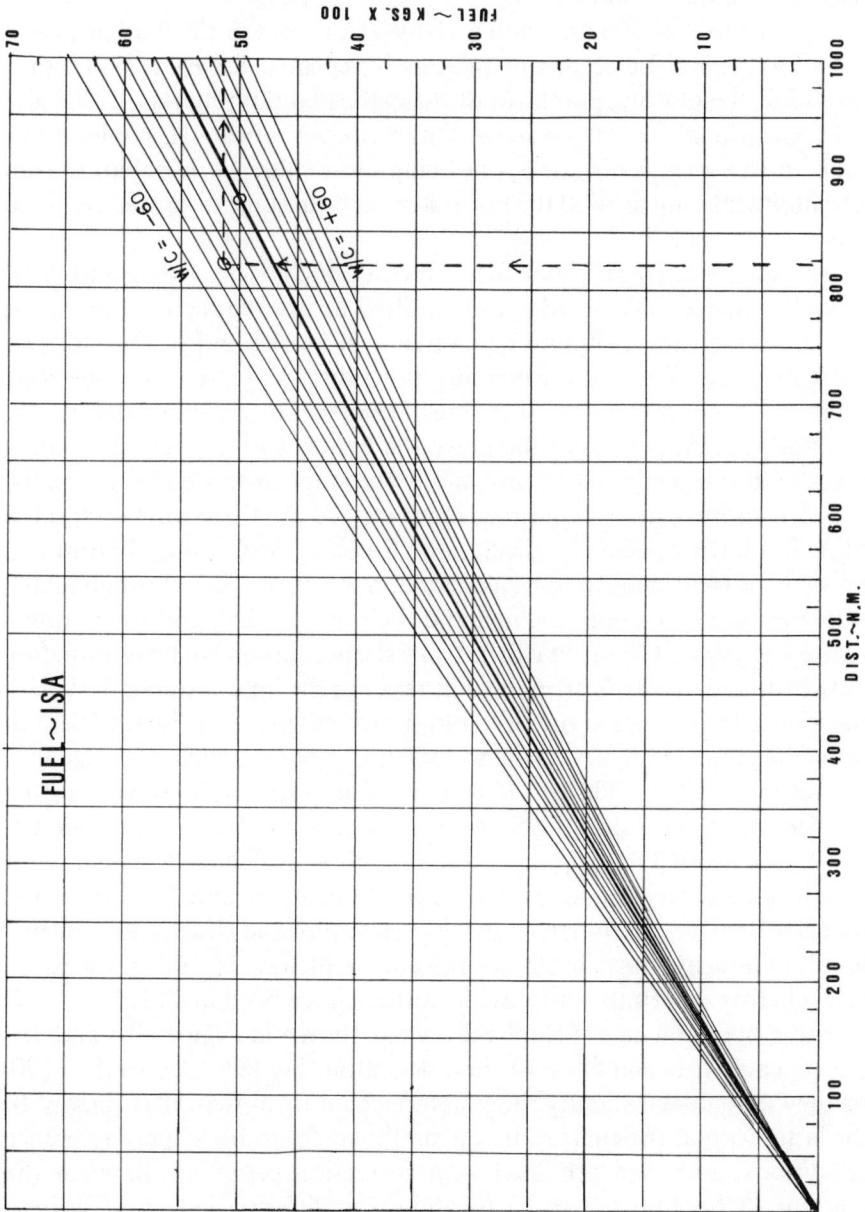

Fig. 7.4   A companion chart to that shown in Fig. 7.3, but this time for fuel burnoff and
including 5% contingency, but nil ground allowance.

minutes while the fuel burned is 2181 kg. So we now add to the cruise (CRZ) columns for time and fuel the climb and descent increments, namely 38 minutes and 1474 kg. This gives us the time and fuel totals for a stage distance of 500 nm, with a –60 W/C; these are 105.0 minutes and 3655 kg. As will be seen, this process is repeated for each W/C, thus providing the plotting points for drawing the planning charts Figs 7.3 and 7.4 appropriate to a 500 nm stage. This process is repeated for at least two more differing stage distances, one being a low value, i.e. less than 100 nm and another being around the normal maximum stage distance, say, 1400 nm.

As explained, contingency fuel is already included; taxying and APU fuel allowances may be added or omitted as desired, as long as this is made clear on the planning chart that will be the end-product of this exercise. Now repeat the foregoing procedure at least twice, selecting arbitrary stage lengths on either side of the 500 nm stage calculated (say, 750 nm and 250 nm – but the more the better, within reason). Using a sheet of graph paper, plot horizontally the stage distance and vertically, fuel. On another sheet, plot time. Thus, for 500 nm there will be a total of thirteen plotting points vertically, one for each W/C value. Repeat this process for each stage length calculated. Now draw a line through all the –60 W/C plots (for both the fuel and the time graphs), and so on, down to the +60 W/C. The result will be a fan-shaped 'family' of lines, radiating outwards and upwards, from zero distance to the limit distance drawn on the graph. In the case of the BAe 146–100 it will probably be sufficient to draw the graph from, say 25 nm to 1500 nm. Now, for each stage distance plotted, e.g. 250 nm, 500 nm, 750 nm, compile a small data box giving the FL for which the calculations were made, e.g. for 500 nm FL 280, the TOW, the mean TAS and the unfactored flow. Calculate for each set of conditions a correction in terms of fuel or time to be added or subtracted for each 1000 ft deviation from the quoted datum and clearly label the fuel graph to indicate that the 5% contingency is included (if this *is* the case), and whether or not the ground allowances have been included.

These planning charts can be drawn as shown in Figures 7.3 and 7.4, but repeated for each 5° or 10° ISA deviation, say ISA –15° to ISA +30, or they may take a slightly more complicated form than that shown. In the latter format the chart is drawn, as above, for a single set of reference conditions, and then provided with correction boxes to allow for the readout to be adjusted so as to allow for differing values of weight, temperature, and FL. This method, although making for a more lengthy chart, and user's time involved, does give a somewhat more accurate result and should be used where the subject aircraft is sensitive. But either method should be suitable for most, if not all, short and medium haul aircraft. It is of interest to note that both types of presentation were used by the author for different variants of the Viscount aircraft, and the latter method was, in fact, later on published by the aircraft's manufacturer. The latter method was also used for the Fokker F27-200 by the author – again successfully. A variation on these two presentations is given in Fig. 7.5, as used by Dan-Air. But please note, *these figures are given for information and illustration purposes only and do not purport to be accurate data and therefore should not be used operationally*.

So far we have dealt only with the departure to destination, or the actual stage. However, as previously mentioned, it is necessary that provision must be made for carrying enough fuel, in addition to the stage fuel, to carry out a diversion to a nominated alternate (or alternates), plus holding fuel sufficient to meet the requirements as laid down for the minimum sector fuel (see Chapter 1). This can conveniently be done by using the same process and format as illustrated in Figs 7.3 and 7.4, but using max. landing weight (MLW) as the starting weight, and also including the contingency allowance. But as this is reserve fuel, having calculated the diversion fuel and time, inclusive of the contingency fuel allowance, add the prescribed holding fuel to the diversion fuel, before drawing the reserve fuel and time charts. (See Fig. 7.6.)

The use of these charts is precisely the same as in the case of the stage fuel and time. The procedure for calculating the required sector fuel before a flight departs is simple, and is as follows:

(1) Obtain the forecast winds, namely the wind velocities (W/V) expressed in degrees true, and speed in knots. Depending on the distance involved there may be anything from one wind velocity to several. Taking the average true track (Av.Tr.(T)) from either the route book, or the plog, as applicable to the stage, calculate for each W/V, from its angle off Tr. (T), a value of wind component. (A table or chart for this purpose, can be contained in the route book in this section, or a navigational computer (pocket version) may be used.) Note each W/C thus obtained for the stage and add these together algebraically, taking care to note the signs + or –. Divide the results (+ and –) by the number of W/C values obtained then add the two totals; this is the stage W/C (see Table 7.7). For example, let us assume that there are three W/C values for a flight. The first is, say –30 kt, the second –15 kt, and the third +20 kt. –30 plus –15 = –45. Adding +20 we get –25. The average W/C is, therefore, $-25 \div 3 = -8.3$ kt or for all practical purposes, –8 kt, this, of course, being a headwind average for the stage.

(2) Repeat the process for the diversion, using the appropriate charts and

forecast W/Vs. Add the stage fuel and the reserve fuel and you have the required sector fuel.

It must not be forgotten that the standard times and MSF quantity, included in the sector details in the route book, plog, or flight brief, will have been calculated using statistical winds, and the appropriate W/C resulting from these will be shown. It is possible that, on occasions, the W/C forecast for the stage and diversion will result in a fuel *requirement* less than MSF. If so, MSF *must* be loaded. And, as regards time, each W/V will have produced its own W/C for a part of the stage and diversion elements. Let us suppose that the *standard* W/C is –2 kt for the stage. But, in (1) above the W/C for the first element is –30 kt. The standard times between the waypoints or RPs affected by this element, must therefore be adjusted to allow for an extra 28 kt headwind on that particular element, while for the next element the standard times are calculated for a 2 kt headwind, whereas the forecast W/C is for a 15 kt headwind. The standard times must, therefore, be adjusted to show 13 kt reduction in the expected ground speed. Adding up the revised times between the various waypoints or RPs will give a revised flight time for the flight plan, as found for the forecast W/C, –8 kt. The same process is also applied to the diversion element, in case it is necessary to divert from the destination. This may, at first sight, appear to be complicated, but in practice it is about a five minute task to carry out. And don't forget to add the taxying allowances, if not included in the charts!

In Fig. 7.5 the Dan-Air Time and Fuel planning chart for ISA + 10° is shown. Unlike the methods described earlier, this allows both parameters to be combined in one chart. Its use is equally simple; enter with stage distance and correct for the W/C. Proceed upwards to the allocated or chosen FL, and move left to obtain the stage fuel. Continue upwards from the FL line and read off the stage time, corrected for the FL to be used. For example, taking the stage distance 700 nm, and with a headwind component of 50 kt, proceed to intersect the FL 290 line. Move left to the fuel scale and then correct for an estimated landing weight of 30 000 kg. Read off the flight fuel for a W/C –50 kt and a LW of 30 000 kg, of 4500 kg. (Were the stage to be flown at FL 190 the flight fuel would increase to 5050 kg.) Now return to the vertical line at the FL 290 point and proceed up to the time grid. It will be seen that, at FL 290 the flight time is 2 hours 18 minutes, increasing to 2 hours 29 minutes at FL 190.

An alternative method of planning sector fuel and time has been evolved by British Aerospace for the BAe 146. This is presented in Table 7.5 and comprises tabulated time and fuel data, for 1000 ft intervals of FL

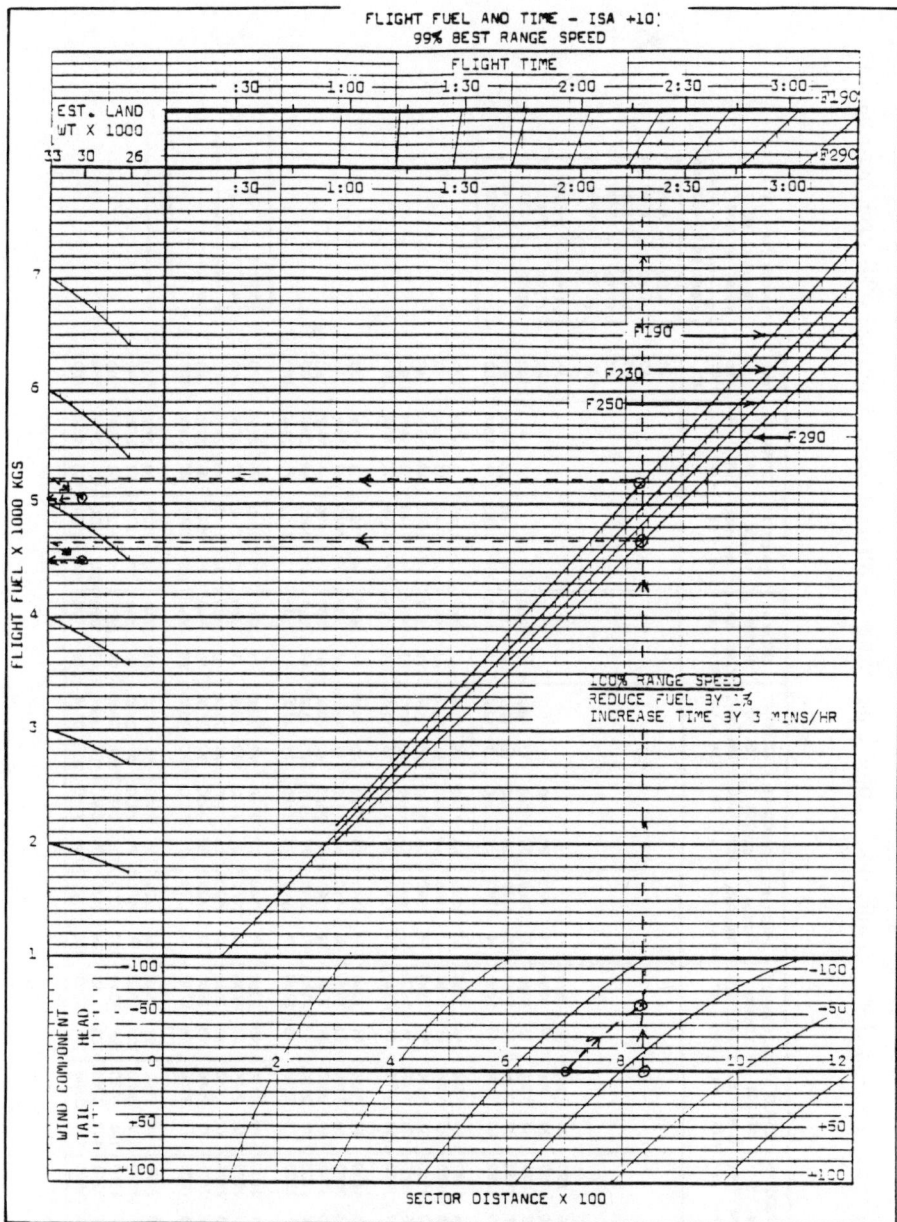

FLIGHT FUEL AND TIME - ISA +10'
99% BEST RANGE SPEED

SECTOR FUEL AND TIME GRAPH

Fig. 7.5   A combined time and fuel chart (ISA +10°) for the BAe 146–100, as used by Dan-Air Services. (Courtesy of British Aerospace and Dan-Air Services Ltd.)

Table 7.5 Flight planning – sector fuel and time. Long range cruise – high alt.:ISA. (Courtesy of British Aerospace.)

APPROX. ALT. AT TOP OF CLIMB SHORT SECTORS	
28000 KG	34000 KG
4600	4200
9400	8500
13700	12300
17700	15700
21100	18800
24200	21500
27200	24400
29700	26300
	28500
	30100

CRUISE ALT(FT)	MAXIMUM T.O.W (KG)
UP TO	
31000	38000

AIR DIST NM	22000 FT TIME MIN	22000 FT FUEL KG	23000 FT TIME MIN	23000 FT FUEL KG	24000 FT TIME MIN	24000 FT FUEL KG	25000 FT TIME MIN	25000 FT FUEL KG	26000 FT TIME MIN	26000 FT FUEL KG	27000 FT TIME MIN	27000 FT FUEL KG	28000 FT TIME MIN	28000 FT FUEL KG	29000 FT TIME MIN	29000 FT FUEL KG	31000 FT TIME MIN	31000 FT FUEL KG	ALT AIR DIST NM
20																			20
40																			40
60																			60
80																			80
100																			100
120	26	880	26	880	26	880													120
140	30	980	29	980	29	980	29	980	29	980									140
160	33	1080	33	1080	33	1080	33	1040	33	1040	33	1060	32	1060					160
180	37	1180	37	1180	36	1160	36	1160	36	1160	36	1160	36	1160	36	1160	36	1160	180
200	41	1280	40	1280	40	1260	40	1260	39	1260	39	1260	39	1260	39	1260	39	1240	200
220	44	1380	44	1380	44	1360	43	1360	43	1360	43	1340	42	1340	42	1340	42	1340	220
240	48	1480	48	1480	47	1460	47	1460	46	1460	46	1440	46	1440	46	1440	45	1420	240
260	52	1580	51	1580	51	1560	50	1560	50	1560	50	1540	49	1520	49	1520	49	1520	260
280	55	1680	55	1680	54	1660	54	1660	53	1660	53	1620	53	1620	52	1620	52	1600	280
300	59	1780	58	1780	58	1760	57	1740	57	1740	56	1720	56	1720	56	1700	55	1700	300
320	63	1880	62	1880	61	1860	61	1840	60	1820	60	1820	59	1800	59	1800	58	1780	320
340	66	2000	66	1980	65	1960	64	1940	64	1920	63	1900	63	1900	62	1880	61	1880	340
360	70	2100	69	2080	69	2060	68	2040	67	2020	67	2000	66	1980	65	1980	65	1960	360
380	74	2200	73	2180	72	2160	71	2140	71	2120	70	2100	69	2080	69	2080	68	2060	380
400	77	2300	77	2280	76	2260	75	2240	74	2220	73	2200	73	2180	72	2140	71	2140	400
420	81	2400	80	2380	79	2360	78	2340	77	2300	77	2280	76	2260	75	2260	74	2240	420
440	85	2520	84	2480	83	2460	82	2440	81	2400	80	2380	79	2360	79	2340	77	2320	440
460	88	2620	87	2580	86	2560	85	2520	84	2500	83	2480	83	2440	82	2440	80	2420	460
480	92	2720	91	2680	90	2660	89	2620	88	2600	87	2580	86	2540	85	2540	84	2520	480
500	96	2820	95	2800	94	2760	92	2720	91	2700	90	2680	89	2640	89	2640	87	2600	500
520	100	2920	98	2900	97	2860	96	2820	95	2800	94	2760	93	2740	92	2720	90	2700	520
540	103	3040	102	3000	101	2960	99	2920	98	2900	97	2860	96	2840	95	2820	93	2780	540
560	107	3140	106	3100	104	3060	103	3020	102	2980	101	2960	99	2920	98	2920	96	2880	560
580	111	3240	109	3200	108	3160	106	3120	105	3080	104	3060	103	3020	102	3000	100	2960	580
600	114	3340	113	3300	111	3260	110	3220	109	3180	107	3160	106	3120	105	3100	103	3060	600

Table 7.5 Continued.

	22000 FT	23000 FT	24000 FT	25000 FT	26000 FT	27000 FT	28000 FT	29000 FT	31000 FT	
620	118 3440	116 3400	115 3360	113 3320	112 3280	111 3240	109 3220	108 3200	106 3160	620
640	122 3560	120 3500	118 3460	117 3420	116 3380	114 3340	113 3300	112 3280	109 3240	640
660	125 3660	124 3600	122 3560	121 3520	119 3480	118 3440	116 3400	115 3380	112 3340	660
680	129 3760	127 3720	126 3660	124 3620	122 3580	121 3540	119 3500	118 3480	116 3420	680
700	133 3860	131 3820	129 3760	128 3720	126 3640	124 3620	123 3580	121 3560	119 3520	700
720	136 3960	135 3920	133 3860	131 3820	129 3760	128 3720	126 3680	125 3660	122 3620	720
740	140 4080	138 4020	136 3960	135 3920	133 3840	131 3820	130 3780	128 3760	125 3700	740
760	144 4180	142 4120	140 4040	138 4020	136 3940	135 3920	133 3880	131 3840	128 3800	760
780	147 4280	145 4220	143 4160	142 4120	140 4040	138 4020	136 3960	135 3940	132 3880	780
800	151 4400	149 4340	147 4280	145 4220	143 4160	141 4120	140 4040	138 4040	135 3980	800
820	155 4500	153 4440	151 4380	149 4320	147 4260	145 4220	143 4160	141 4140	138 4080	820
840	158 4620	156 4540	154 4480	152 4420	150 4360	148 4300	146 4260	145 4220	141 4160	840
860	162 4720	160 4640	158 4580	156 4520	154 4460	152 4400	150 4360	148 4320	144 4260	860
880	166 4820	164 4740	161 4680	159 4620	157 4560	155 4500	153 4460	151 4420	147 4360	880
900	169 4940	167 4860	165 4780	163 4720	160 4660	158 4600	156 4540	154 4520	151 4460	900
920	173 5040	171 4960	168 4900	166 4820	164 4760	162 4700	160 4640	158 4620	154 4540	920
940	177 5140	174 5080	172 5000	170 4920	167 4860	165 4800	163 4740	161 4720	157 4640	940
960	181 5260	178 5180	176 5100	173 5040	171 4960	169 4900	167 4840	164 4800	160 4740	960
980	184 5360	182 5280	179 5200	177 5140	174 5060	172 5000	170 4940	168 4900	163 4840	980
1000	188 5460	185 5380	183 5300	180 5240	178 5160	175 5100	173 5040	171 5000	167 4920	1000
1020	192 5580	189 5500	186 5400	184 5340	181 5260	179 5200	176 5140	174 5100	170 5020	1020
1040	195 5680	193 5600	190 5520	187 5440	185 5360	182 5300	180 5220	177 5200	173 5120	1040
1060	199 5780	196 5700	193 5620	191 5540	188 5440	186 5400	183 5320	181 5280	176 5220	1060
1080	203 5900	200 5820	197 5720	194 5640	192 5560	189 5500	186 5420	184 5380	179 5300	1080
1100	206 6000	203 5920	201 5820	198 5740	195 5660	192 5600	190 5520	187 5480	183 5400	1100
1120	210 6120	207 6020	204 5920	201 5840	198 5760	196 5680	193 5620	191 5580	186 5500	1120
1140	214 6220	211 6120	208 6040	205 5940	202 5860	199 5780	196 5720	194 5680	189 5600	1140
1160	217 6340	214 6240	211 6140	208 6040	205 5960	203 5880	200 5820	197 5780	192 5680	1160
1180	221 6440	218 6340	215 6240	212 6160	209 6060	206 5980	203 5900	201 5860	195 5780	1180
1200	225 6560	222 6440	218 6360	215 6260	212 6160	209 6080	207 6000	204 5960	199 5880	1200
1220	228 6660	225 6560	222 6460	219 6360	216 6280	213 6180	210 6100	207 6060	202 5980	1220
1240	232 6780	229 6660	226 6560	222 6440	219 6380	216 6300	213 6200	210 6160	205 6080	1240
1260	236 6880	232 6780	229 6680	226 6580	223 6480	220 6400	217 6300	214 6260	208 6160	1260
1280	239 7000	236 6880	233 6780	229 6680	226 6580	223 6500	220 6400	217 6360	211 6260	1280
1300	243 7100	240 7000	236 6880	233 6780	230 6680	226 6600	223 6500	220 6460	215 6360	1300

CORRECTION TO TABLES

Table 7.5 Continued.

	22000 FT	23000 FT	24000 FT	25000 FT	26000 FT	27000 FT	28000 FT	29000 FT	31000 FT
1320	247 7240	243 7100	240 7000	236 6880	233 6780	230 6700	227 6620	224 6560	218 6440
1340	251 7320	247 7220	243 7100	240 7000	236 6900	233 6800	230 6720	227 6640	221 6560
1360	254 7440	251 7320	247 7200	243 7100	240 7000	237 6900	233 6820	230 6760	224 6660
1380	258 7540	254 7440	250 7320	247 7200	243 7100	240 7000	237 6920	234 6860	227 6760
1400	262 7640	258 7540	254 7420	250 7320	247 7200	243 7100	240 7020	237 6960	231 6860
1420	265 7740	261 7640	258 7520	254 7420	250 7300	247 7200	243 7120	240 7060	234 6960
1440	269 7840	265 7760	261 7640	257 7520	254 7400	250 7320	247 7220	243 7160	237 7060
1460	273 8000	269 7860	265 7740	261 7620	257 7520	254 7420	253 7320	247 7260	240 7160
1480	276 8100	272 7980	268 7840	265 7740	261 7620	257 7520	253 7420	250 7360	243 7260
1500	280 8220	276 8080	272 7960	268 7840	264 7720	260 7620	257 7520	253 7460	246 7360
1520	284 8340	280 8200	275 8060	272 7940	268 7820	264 7720	260 7620	257 7560	250 7460
1540	287 8440	283 8300	279 8180	275 8040	271 7920	267 7820	263 7720	260 7660	253 7560
1560	291 8560	287 8420	283 8280	279 8160	275 8020	271 7920	267 7820	263 7760	256 7640
1580	295 8680	290 8540	286 8400	282 8260	278 8140	274 8020	270 7920	267 7860	259 7740
1600	298 8900	294 8640	290 8500	286 8380	281 8240	277 8140	273 8020	270 7960	262 7840

CORRECTION TO SECTOR TIME FOR ARRIVAL WEIGHT

SECTOR TIME (MIN) FROM MAIN TABLE	50	100	150	200	250
ARRIVAL WEIGHT (KG):					
28000	0	0	0	0	0
29000	0	0	0	0	0
30000	0	0	0	0	0
31000	0	0	0	0	+1
32000	0	0	0	+1	+1
33000	0	+1	+1	+1	+1
34000	+1	+1	+1	+1	+1

CORRECTION TO SECTOR FUEL FOR ARRIVAL WEIGHT

SECTOR FUEL (KG) FROM MAIN TABLE	2000	3000	4000	5000	6000	7000	8000
ARVL WT (KG):							
28000	0	0	0	0	0	0	0
29000	+50	+50	+50	+50	+100	+100	+100
30000	+100	+100	+100	+100	+150	+200	+200
31000	+100	+150	+150	+200	+250	+300	+350
32000	+150	+200	+200	+250	+300	+400	+400
33000	+200	+250	+300	+350	+400	+450	+500
34000	+250	+300	+350	+450	+450	+550	+450

Table 7.6 Alternate fuel and trim tabulations. (Courtesy of Dan-Air Services Ltd.)

WIND:		-50	-40	-30	-20	-10	0	+10	+20	+30	+40	+50	
Dist:	20.	332.	319.	305.	292.	278.	265.	256.	250.	250.	250.	250.	Dst: 20.
Time:		8.	8.	7.	6.	6.	6.	5.	6.	6.	6.	6.	
Level:		7.	6.	6.	5.	4.	3.	3.	0.	0.	0.	0.	
Dist:	25.	355.	341.	328.	314.	301.	287.	273.	259.	250.	251.	250.	Dst: 25.
Time:		10.	9.	8.	7.	7.	6.	6.	6.	6.	6.	6.	
Level:		9.	8.	7.	6.	5.	5.	4.	3.	3.	0.	0.	
Dist:	30.	408.	387.	366.	345.	324.	303.	289.	275.	261.	250.	250.	Dst: 30.
Time:		11.	10.	9.	9.	9.	7.	7.	6.	6.	6.	6.	
Level:		9.	8.	8.	7.	5.	5.	5.	4.	3.	3.	0.	
Dist:	35.	443.	422.	401.	380.	359.	339.	324.	310.	296.	282.	268.	Dst: 35.
Time:		12.	12.	11.	10.	9.	9.	8.	7.	7.	6.	6.	
Level:		10.	10.	9.	8.	7.	6.	6.	5.	5.	4.	4.	
Dist:	40.	478.	457.	436.	415.	394.	373.	359.	345.	331.	317.	303.	Dst: 40.
Time:		13.	13.	12.	11.	11.	10.	9.	9.	8.	8.	7.	
Level:		10.	10.	10.	9.	9.	8.	7.	7.	6.	6.	5.	
Dist:	45.	513.	492.	471.	450.	429.	408.	394.	380.	366.	352.	338.	Dst: 45.
Time:		15.	14.	13.	13.	12.	11.	11.	10.	9.	9.	8.	
Level:		11.	11.	10.	10.	10.	9.	9.	8.	8.	7.	6.	
Dist:	50.	530.	512.	494.	476.	458.	440.	428.	416.	404.	392.	380.	Dst: 50.
Time:		16.	15.	14.	14.	13.	12.	12.	11.	11.	10.	10.	
Level:		12.	11.	11.	11.	10.	10.	10.	10.	9.	8.	8.	
Dist:	55.	590.	566.	542.	518.	494.	470.	452.	434.	416.	398.	380.	Dst: 55.
Time:		17.	17.	16.	15.	14.	13.	13.	12.	11.	11.	10.	
Level:		13.	12.	12.	11.	11.	10.	10.	10.	9.	8.	9.	
Dist:	60.	620.	596.	572.	548.	524.	500.	482.	464.	446.	428.	410.	Dst: 60.
Time:		18.	17.	17.	16.	16.	15.	14.	13.	13.	12.	11.	
Level:		13.	13.	12.	12.	11.	11.	10.	10.	10.	10.	9.	
Dist:	65.	650.	626.	602.	578.	554.	530.	512.	494.	476.	458.	440.	Dst: 65.
Time:		19.	18.	18.	17.	16.	16.	15.	14.	14.	13.	12.	
Level:		14.	13.	13.	12.	12.	11.	11.	11.	10.	10.	10.	
Dist:	70.	680.	656.	632.	608.	584.	560.	542.	524.	506.	488.	470.	Dst: 70.
Time:		20.	19.	19.	18.	17.	16.	16.	16.	15.	14.	13.	
Level:		14.	14.	13.	13.	12.	12.	12.	11.	11.	10.	10.	
Dist:	75.	710.	686.	662.	638.	614.	590.	572.	554.	536.	518.	500.	Dst: 75.
Time:		22.	20.	19.	19.	18.	17.	17.	16.	16.	15.	15.	
Level:		15.	14.	14.	13.	13.	13.	12.	12.	11.	11.	11.	
Dist:	80.	740.	716.	692.	668.	644.	620.	602.	584.	566.	548.	530.	Dst: 80.
Time:		23.	22.	20.	20.	19.	18.	18.	17.	17.	16.	16.	
Level:		16.	15.	15.	14.	14.	13.	13.	12.	12.	12.	11.	
Dist:	85.	770.	746.	722.	698.	674.	650.	632.	614.	596.	578.	560.	Dst: 85.
Time:		24.	23.	22.	21.	20.	19.	19.	18.	17.	17.	16.	
Level:		16.	16.	15.	15.	14.	14.	13.	13.	13.	12.	12.	
Dist:	90.	800.	776.	752.	728.	704.	680.	662.	644.	626.	608.	590.	Dst: 90.
Time:		25.	24.	23.	22.	22.	20.	19.	19.	18.	18.	17.	
Level:		17.	17.	16.	15.	15.	14.	14.	14.	13.	13.	13.	
Dist:	95.	830.	806.	782.	758.	734.	710.	692.	674.	656.	638.	620.	Dst: 95.
Time:		25.	25.	24.	23.	23.	22.	20.	20.	19.	19.	18.	
Level:		18.	17.	17.	16.	15.	15.	15.	14.	14.	13.	13.	
Dist:	100.	848.	826.	804.	783.	761.	740.	723.	707.	691.	675.	659.	Dst: 100.
Time:		26.	26.	25.	24.	23.	23.	22.	22.	20.	20.	19.	
Level:		19.	18.	18.	17.	16.	16.	15.	15.	14.	14.	14.	
Dist:	105.	902.	875.	848.	821.	794.	767.	745.	723.	702.	680.	659.	Dst: 105.
Time:		28.	27.	26.	25.	25.	24.	23.	22.	21.	20.	19.	
Level:		20.	19.	18.	18.	17.	17.	16.	15.	15.	14.	14.	
Dist:	110.	929.	902.	875.	848.	821.	794.	772.	750.	728.	707.	686.	Dst: 110.
Time:		29.	28.	27.	26.	25.	25.	24.	23.	22.	22.	20.	
Level:		21.	20.	19.	19.	18.	17.	16.	16.	15.	15.	14.	
Dist:	115.	956.	929.	902.	875.	848.	821.	799.	777.	755.	734.	713.	Dst: 115.
Time:		30.	29.	28.	27.	26.	25.	25.	24.	23.	23.	22.	
Level:		22.	21.	20.	19.	18.	18.	17.	17.	16.	15.	15.	

ALTERNATE FUEL AND TRIM TABULATIONS

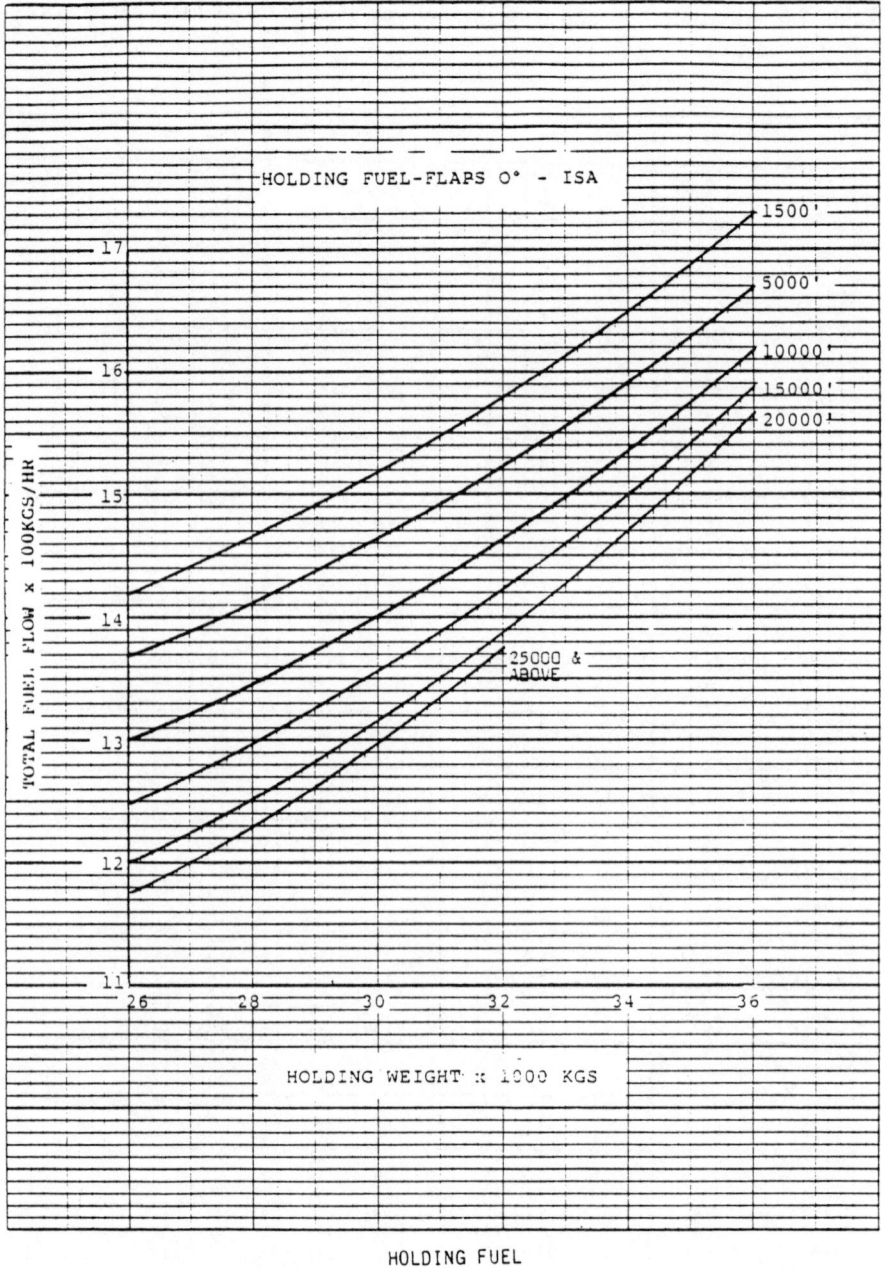

Fig. 7.6   A simple chart showing holding fuel flow in kg/hr for the BAe 146–100, as used by Dan-Air Services. (Courtesy of British Aerospace and Dan-Air Services Ltd.)

and 20 mile intervals as regards distance. But is must be noted that the distance values shown are in *air* nautical miles. In other words, the route wind component must be applied to the still air distance in order to obtain the air distance. Each page of tabulation is for a specific ISA temperature deviation: Table 7.5 is given for ISA. Correction boxes are provided to allow for both the sector time and fuel values extracted from the table to be corrected backwards from expected arrival weight. (A method of obtaining the Route W/C is shown in Table 7.7.)

To calculate alternate fuel, Dan-Air use a tabular presentation, as shown in Table 7.6. Here the diversion distance is tabulated at 5 nm intervals, and a value of alternate fuel burnoff, time, and optimum FL is given, as appropriate to the forecast diversion W/C. Holding fuel is presented graphically (see Fig. 7.6), and must be added to the alternate fuel obtained from Table 7.6. If reference is made to Tables 7.2, 7.3, and 7.4 it can readily be seen how much time is saved by using the precalculated data discussed. See also Table 7.7. (For example Dan-Air use a standard holding procedure, using 30 minutes at 1500 ft, ISA +10°C at MLW, which gives a value of 840 kg.) But the quantity of holding fuel to be loaded must depend upon the alternate's characteristics. If, for example, the destination is an ocean island – such as Ascension Island – the minimum acceptable holding time will be 2 hours, there being no suitable alternate within acceptable range. The forecast meteorological conditions would also have to be taken into account. In such cases an acceptable procedure involving holding altitude would have to be evolved, and the chart forming Fig. 7.6 used to obtain the quantity of fuel to hold for 2 hours, at the weight and altitude applicable.

Table 7.7 Calculation of route wind component.

Wind component Av. Tr.(T) = 123°		
	W/C	
W/V	+	–
090/25		20
179/30		19
270/33	29	
Averages	29	39
Route W/C	-3	

Finally, a commercial matter that has been briefly mentioned earlier in this book, although it should be taken into account in all preflight procedures. It can be of significant importance in keeping stage costs to a minimum, or, to use another expression, to help keep these costs to a level that need not be exceeded unnecessarily.

Fuel Price Index

Fuel costs are one of the highest of the hourly costs incurred in flying these days, and can be *the* highest. It can be a very useful function of route planning to estimate reasonably precisely the annual uplift on each and every stage that the airlines flies, and to keep these estimates separate for each stage. Annually these estimated requirements are then communicated to the various fuel suppliers involved at each airport, and a contract price obtained where possible for the unit of measure used. Each price will be valid only for a specific airport, and should be treated as being commercially confidential. Quite understandably there will be a very wide variation in price from airport to airport, depending on a number of factors, such as the distance that the fuel has to be transported from the refinery, currency conversion rates, and so on. (In some regions prices will fluctuate so frequently that it becomes only possible to provide this information outside the route book, such as on the crew-room notice board, where it can be more easily kept up to date.)

Because of the commercial nature of the information, and the possible help that possession of this information might be to a competitor, it is not desirable that actual prices should be given in the route book, nor communicated to crews. At the same time it is most desirable that the commander of an airliner should have some guidance as to the *relationship* between these prices, using a single common denominator. Given such information it can then be decided by a commander whether or not he should load fuel in excess of that required for a stage, subject, of course, to (a) whether his takeoff weight will permit this, and (b) he has the available tank capacity. It must also be borne in mind that the simple fact of carrying excess fuel can cost money due to the operational implications. For example, extra all-up weight can involve a reduction in TAS, and therefore increased stage time and cost. It can also affect the flight level available or attainable, and this may involve an increase in fuel flow and consequently burnoff. Thus before loading excess fuel there must exist a sufficient differential to make this worthwhile and some guidance must be given as to the cost below datum above which excess fuel should not be loaded.

One simple method of presenting this data is to provide a simple table of all airports regularly visited, each with an *indication* as to the cost of fuel, per unit. The airline's base can conveniently form the datum price, and this can be shown as 100(%). Taking all other airports in the table in alphabetical order, convert the contract price into a percentage of the base, or datum price; a suitable layout is given in Table 7.8, in a very short form. There is no significance in the figures quoted and they are simply illustrations as to layout.

Table 7.8 Fuel price index.

Airport	Index	Supplier
Gatwick (Base)	100	Shell
Auckland	125	Esso
Bahrein	70	Aramco
Cairo	83	Misroil
Düsseldorf	90	Mobil
Edinburgh	105	BP
Faro	115	Shell
Geneva	120	Total
Hamburg	90	Texaco
Istanbul	120	Gulf
Jersey	125	Shell
Khartoum	170	Shell
Lisbon	105	Esso
Manchester	98	BP

Do not uplift 'Economy fuel' unless the index is below 90

The suppliers listed may, or may not, be at the airport, or even in existence at the time of writing.

8: Aerodrome Operating Minima

One vital, but most uninspiring, content of any route book, or flight brief, is a tabulation of significant meteorological conditions, in terms of visibility and cloud base, below which an operator's aircraft may not land or takeoff. This is, normally, a legal requirement and woe betide the pilot that is caught breaking the rule! The penalties can include the withdrawal of his, or her, professional licence, and this can, of course, lead to the loss of employment. As regards the expression above, 'caught breaking the rule', do not think that, because a landing has been made in conditions below those specified, successfully, all is well. It is agreed that it *probably* is, but it should not be forgotten that the airport always logs the time of landing, together with regularly updated reports of the actual meteorological conditions obtaining. Thus, if the authority wishes to carry out a random 'spot' check, it has the means of doing this. In this chapter the elements of aerodrome operating minima (AOM) will be presented, together with some indication as to how they are calculated.

We are dealing with a complex subject, and one that has variations according to the regulatory body involved. Basically, AOM is derived from the use of a radio navigational aid, associated with the highest obstacle within a designated area for that aid. This obstacle controls the lowest height or altitude, to which an aircraft may descend on an approach to land under instrument flight rules. Every radio aid type has a different defined obstacle area; the simpler the aid the larger the area. So, for a non-directional beacon (NDB), because of its relative inaccuracy, the defined area for obstacle clearance purposes is substantially greater than that for a precision aid, such as the instrument landing system (ILS). The geographical location of the aid must also be taken into account. For example, unless it is in line with the runway involved, that is to say, is located on the runway's extended centre line, it cannot be used as an approach aid, and is restricted for use to descending below cloud ('cloud break'), after which the aircraft must manoeuvre visually to line-up with the runway. Another significant factor is the type of approach lighting and runway lighting that is available and this controls the minimum acceptable visibility. (Please note that, at present, we are still using something close to lay language, in the interests of clarity.) Thus, the height of the significant obstacle controls the lowest height to which an aircraft may descent using a particular aid, while the type of approach and runway lighting controls the required horizontal visibility at this height. The better the lighting pattern (GGP) the lower the required visibility. And finally, before we move on to detail, the aircraft's 'category' also affects the minimum values of vertical and horizontal elements. There are two systems of categorising airliners, and these are (a) weight related and (b) speed related. The former divides aircraft into a number of weight groupings, while the second uses approach speed groupings. The former method is currently being withdrawn, progressively, and this will eventually leave speed related groupings as the sole criteria.

UK CAA Method

See Appendix B for UK Air Navigation Order definitions.

Let us now look briefly at the methods of arriving at AOM in general use. First, the UK Civil Aviation Authority method, as contained in the publication, CAP 360. This is now in the process of being replaced by the International Civil Aviation Organisation's method. CAP 360 is a weight related method and relies on initially establishing a decision height (DH) for the aid being used in a landing. For this aid an obstacle clearance limit (OCL) is published, and a prescribed increment must be added to this value in order that a DH value be derived. This increment varies with the weight category of the aircraft, so that aircraft in different weight brackets will have different values of DH. DH values are related to the elevation of the runway threshold, except for cloud break (CB) and circling, when the elevation of the airport itself is used. Allowance must also be made for altimeter position error, after which we have a value of DH that can be published. Using this DH, we can then obtain an associated visibility value, known as runway visual range or RVR. RVR is obtained simply by entering a table with the DH and reading off the required RVR as appropriate to the approach and runway lighting pattern and characteristics. (We will examine this method of deriving AOM later in the chapter.) It is a relatively simple method, but will soon become defunct in the interests of international standardisation, although the RVR table will remain, for the time being.

It should perhaps be mentioned that the expression aerodrome operating minima is *relatively* new, and was doubtless adopted under the general term 'simplification'. Its previous title was weather minima (WXM), and this can clearly be seen to be a comparatively cumbersome title! Similarly, under the same process of simplification, came the changes in the method of calculation (although it must be said that CAP 360 at least is now forthcoming). In the UK, the accepted method of calculating weather minima was quite simple. One took the published

OCL, rounded it up to the nearest 10 ft, and thus derived a DH. RVR was then obtained by the simple formula

$$\frac{V_{AT_0}}{60} \times DH$$

and this gave a value of RVR. It was not normal for any method to be published, or approved, and the operator simply submitted his weather minima to the then regulatory body, for 'approval'. It used to be said, perhaps unfairly, that when the regulatory body ceased objecting this could be constituted as being approval. So the CAP 360 method is a considerable step forward. It should be made clear here that the 'old' method of RVR calculation also involved taking into account the GGP lighting; the formula above was valid only for a full 5-bar Calvert system. If the GGP was less then the RVR value had to be increased on a scale relative to the GGP available.

ICAO Method
As mentioned earlier, the UK CAA method for calculating AOM is being phased out, in favour of the ICAO system. This method, in particular, uses a different system for categorising aircraft groups, basing these on approach speed rather than on weight. So, to some extent, there is a return to the pre-CAP 360 method. It seems far more logical and important, to the author, to link visibility requirements to speed rather than to weight. But the use of the weight classification could be broadly held to involve speed through the function of weight and V_S and thus to V_{AT}, this value being $1.3 \times V_S$ in the landing configuration at the threshold of the runway. However, in the UK the method of calculating the RVR still makes use of the CAP 360 table referred to above, instead of the ICAO method as 'defined' in ICAO PANS/OPS 8168-OPS 611. The latter still remains to be finalised at the time of writing. The detailed methods published by ICAO and the UK variant will be gone into later.

Finally, in this generalised preamble, it should also be mentioned that some states publish their own AOM, and such values must be observed. However, an operator may submit alternative minima to the state involved with a request that this may be specifically approved for that operator, i.e. the operator requests that its aircraft may be permitted to operate to its proposed minima, as opposed to the state minima. If approval is forthcoming it is specific to that particular operator. In addition to state minima, some states also vary the ICAO method in certain respects, particularly with regard to the derivation of RVR. The whole of the AOM derivation process, irrespective of methodology is complex and is a veritable morass. As mentioned above, there are three methods for yielding AOM, one of which is dying, the second being subject to national modifications, while the third ignores both but issues its own specific minima – which can be altered by the process of negotiation. The net result is that airlines using the same airport or airports, can be operating, quite legally and properly, to different AOM standards.

Note
Where the expression 'published' is used in the foregoing paragraphs, and in the context of AOM generally, this is intended to mean that the regulatory authority involved has promulgated certain values, such as OCH, OCL, or state minima, by means of an official, or approved document. Normally the state aeronautical information publication (AIP) is the vehicle for this purpose, but other documents, such as AERAD or Jeppesen Flight Guides, may be deemed to be 'approved' for this purpose.

Derivation of AOM – CAP 360 (UK Method)
As has been mentioned earlier, there are precision and non-precision aids to making an instrument landing, and the rules vary for these two groupings. But, first of all, we must become familiar with the current UK CAA aircraft classification or grouping process. In the UK this process is based on maximum landing weight authorised (MLWA), and in this book we are going to ignore aircraft of less than 5700 kg max. TOW. Category A aircraft have a MLWA of less than 25 000 kg. Category B comprises aircraft having a MLWA of between 25 000 kg up to 160 000 kg, while category C have a MLWA in excess of 160 000 kg. It should be noted that AERAD has agreed with the UK CAA that category B may be subdivided into categories B1 and B2 so as to fall more or less into other international categorisations, which have a subdivision between 70 000 kg and 80 000 kg. (France has had weight related categorisation for many years.) Without this subcategorisation, UK aircraft flying in certain states abroad could be heavily penalised by foreign state AOM, e.g. take an aircraft of, say, 70 000 kg flying to an airport having state minima and with a subdivision in the range mentioned above. Under such circumstances, unless the UK category is subdivided, the aircraft would need to comply with the requirements for 160 000 kg. Therefore the CAA has agreed with AERAD that category B may be subdivided so that B1 ranges from 25 000 kg to 68 000 kg MLWA, and B2 from over 68 000 kg to 160 000 kg MLWA.

Precision Approaches, UK
Having established the appropriate category the next thing to be done is

to obtain values of AOM for the aids available at each airport to be used, and this means dividing these into precision and non-precision aids, and also to establish any aids that can only be used for cloud break, due to their not being located in line with the landing runway. Precision aids are instrument landing system (ILS) with glide path (G/P), and precision approach radar (PAR). There is also the microwave landing system (MLS) although this is still something of a rarity, as yet. According to CAP 360 the OCL published for these precision aids must be subject to the addition of a height increment according to the weight category. Thus, taking only transport category aircraft, i.e. those exceeding 5700 kg MTWA – the published OCL must be added as follows:

Cat. A (MLWA exceeding 5700 kg but less than 25 000 kg)   OCL + 20 ft
Cat. B (MLWA 25 000 kg to 160 000 kg)   OCL + 35 ft
Cat. C (MLW exceeding 160 000 kg)   OCL + 50 ft

These increments assume that the ILS G/P is between 2.5° and 3.5°. For G/Ps between 3.6° and 4.0°, the increments are 25 ft, 45 ft, and 70 ft, to be added to the OCL. In all cases, the addition of the OCL and increment produces the decision height (DH) in feet, above the appropriate threshold. In the case where the ILS localiser is offset from the runway centre line the DH may not be less than the height of the G/P at the intersection of the localiser and the runway extended centre line.

Note
CAP 360 does not give any guidance as regards G/P value in excess of 3.5°, and the figures quoted above for steeper values are derived from AERAD, but with a caveat regarding acceptability to a particular aircraft type.

Non-Precision Approaches, UK
For non-precision approaches, the DH may not be less than the OCL published, for the aid being used. While this may, at first sight, appear to be somewhat strange it must be remembered that the defined area for precision approaches is considerably smaller than for other types, and, as will be seen later, the absolute limits are much less restricting. An important point that must be remembered at all times in the course of dealing with AOM is that an ILS approach, without G/P or with G/P inoperative, is *a non-precision approach*. In a full ILS, the localiser (LLZ) provides track guidance (i.e. horizontal guidance) and the G/P gives descent path guidance (i.e. vertical guidance). Thus, ILS (G/P Inop.) provides only one axis of guidance, although the addition of aids such as

distance measuring equipment (DME) assists in providing means for height guidance when established on the LLZ.

Note
In all cases, the DH obtained must be further corrected so as to allow for altimeter position error. The necessary correction can be found from the AFM and after taking this into account, a final value of DH becomes available.

The UK (CAP 360) *minimum* DH values are as follows:

	(above runway threshold)
ILS	200 ft
PAR	200 ft
SRA (search radar) term. $\frac{1}{2}$ nm	250 ft
SRA term. 2 nm	350 ft
VOR (VHF omnidirectional radio range)	300 ft
NDB (non-directional radio beacon)	300 ft
VDF (VHF direction finder)	300 ft
ILS G/P Inop.	250 ft
ILS back beam	250 ft

Note
ILS back beam utilises the reciprocal transmission of the LLZ. The indications will therefore be *reversed* on the aircraft's instrument, i.e. from command to deviation indication.

Circling, UK
Where an aid, such as VOR or NDB, is situated off the extended runway centre line this can only be used for a cloud break procedure. That is to say, a procedure that will bring an aircraft into sight of the airport rather than a specific runway. After breaking cloud into visual meteorological conditions (VMC) the aircraft must then manoeuvre visually to line up with the extended centre line of the runway in use for landing. The height may not be less than 500 ft above airport elevation *or* circling height, whichever is the greater. But, in the case of 'wide-bodied aircraft', the 500 ft minimum height is replaced by 800 ft. (Strictly speaking, circling height should be referred to as visual manoeuvring height.) Associated with this height is a required minimum visibility, in flight visibility (IFV), and this is simply 20 × circling speed for the aircraft type involved. The value is expressed in metres.

Circling minima (for weight related categories) is as follows, and is an absolute minima:

CAP 360 group	Min. obstacle clearance	Min. circling height (above airport)
A	350 ft	500 ft
B (B1–UK)	350 ft	500 ft
(B2–UK)	350 ft	500 ft
C	500 ft	800 ft

Associated with the above is a defined area, within which the significant obstacle lies. In the case of Group A aircraft, this is a circle with a radius of 4 nm from the published airport reference point, while for all other groups the radius is 5 nm. Some states will publish a value of IFV for an indirect approach to a runway, and in such cases the state values apply. However, as regards UK registered aircraft, *the absolute figures shown above must apply even if the state figures are less.* In the UK, the CAA recommends that a minimum RVR of 800 m is used following an indirect approach *or* the lowest listed value of RVR (applicable to Category 1), whichever is the lesser, as applicable to the runway to be used. No cognisance is taken of the GGP or lighting, and all values of DH are rounded *up* to the nearest 50 ft increment. (The Category 1 referred to above refers to a classification of minimum DH/RVR, this being 200 ft/ 600 m. Categories 2 and 3 require special avionics and will be discussed later in the chapter.)

Visibility and RVR, UK
As has been mentioned earlier, the DH is simply the OCL published plus the increments as appropriate to each weight classification. This, however, may not be of a lower value (CAP 360) than the absolute minima DH values quoted earlier, nor the state minima. Although internationally (unless expressly stated to the contrary) the maximum value of RVR permitted is 2000 m, the UK CAA specifies a lower limit, namely 1500 m. However, another state's RVR, up to 2000 m, may be used where applicable. RVR is not necessarily reported at all airports, and visibility will then be reported as meteorological visibility. CAP 360 requires that met. visibility be converted to an equivalent RVR value according to the GGP, and day or night conditions, thus, RVR = met. visibility when:

HI GGP and runway lighting	met. visibility × 1.5 (day) or 2.0 (night)
Any other lighting	met. visibility × 1.0 (day) or 1.5 (night)
No lighting	met. visibility × 1.0 (day) —

This conversion may not necessarily be applicable to other states (e.g. Australia, France, USA and USSR, to name just a few). As regards UK aircraft, in any case where RVR is not reported an operator must specify a minimum value of reported *visibility* below which an approach may not be initiated. Such visibility may be derived from the RVR *required* and adjusted so as to produce a visibility equivalent, as shown above. The RVR required is, of course, applicable to the procedure (see Note 2).

Note
(1) *The term decision height (DH) has been used above. Strictly speaking this term is only applicable to precision approaches, and for non-precision approaches the appropriate term is minimum decision height (MDH).* DH *only* is used by the UK (CAP 360).
(2) The UK approach ban prohibits, by law, a descent below 1000 ft above the airport height if the RVR at the time is less than that specified in the operator's AOM *or* a continued approach below the specified DH or MDH, unless from such height a visual reference for landing has been made and maintained.

In all cases when a value of DH has been obtained, as described in the foregoing pages, the relevant RVR must be established. Under UK regulations this is obtained from Table 8.1, by entering the table with the calculated DH/MDH and moving horizontally to the appropriate column for the GGP available. As an example, assume that the DH for a procedure is 312 ft, and that there are 690 m of high intensity (HI) approach lighting (GGP) available. Table 8.1 shows that the DH/MDH entry is made at 288–325 ft, and that the applicable column is that headed 749–650 m HI lighting. The resulting RVR value is 1000 m.

Takeoff Minima
Take off minima must be specified by the operator, and, in the UK, is not laid down. Cognisance must be taken of the runway lighting and dimensions. Where a state publishes takeoff minima this must be observed, unless that state's approval has been obtained for an operator's own minima.

ICAO PANS-OPS Doc 8168/611 Method
See Appendix B for ICAO definitions.

Having briefly discussed the UK CAA AOM basis, let us now turn our attention to the ICAO method, as described in PANS-OPS 8168/611. Volume I is sufficient for our purposes but for more detail see Volume II. It will be recalled that mention was made earlier of the UK formula for arriving at both DH and RVR, the latter being a function of approach speed. This was a logical way of arriving at minima, speed being the more

Table 8.1 Derivation of RVR from DH/MDH and GGP. (Courtesy of British Airways AERAD.)

Ground Guidance Pattern

High Intensity Approach Lighting
Length Available (Metres) Day/Night

DH/MDH (feet)	Over 850	850 -750	749 -650	649 -550	549 -450	449 350	349 -250	249 -150	149 -0
200-212	600	600	700	700	800	800	900	1000	1100
213-237	600	700	700	800	800	900	1000	1100	1200
238-262	700	700	800	900	900	1000	1100	1200	1300
263-287	700	800	900	1000	1000	1100	1200	1300	1400
288-325	800	900	1000	1100	1100	1200	1300	1400	1500
326-375	900	1000	1100	1200	1200	1300	1400	1400	1500
376-425	1000	1100	1200	1300	1300	1400	1500	1500	1500
426-475	1100	1200	1300	1400	1400	1500	1500	1500	1500
476-525	1200	1300	1400	1500	1500	1500	1500	1500	1500
526-575	1300	1400	1500	1500	1500	1500	1500	1500	1500
576-625	1400	1500	1500	1500	1500	1500	1500	1500	1500
626 or higher	1500	1500	1500	1500	1500	1500	1500	1500	1500

———————— RVR (m) ————————

Low Intensity Approach Lighting
Length Available (Metres)

DH/MDH (feet)	Day All Lengths	Over 600	Night 599 -300	299 -0
200-212	1100	900	1000	1100
213-237	1200	900	1100	1200
238-262	1300	1000	1200	1300
263-287	1400	1100	1300	1400
288-325	1500	1200	1400	1500
326-375	1500	1200	1400	1500
376-425	1500	1300	1500	1500
426-475	1500	1400	1500	1500
476-525	1500	1500	1500	1500
526-575	1500	1500	1500	1500
576-625	1500	1500	1500	1500
626 or higher	1500	1500	1500	1500

———————— RVR (m) ————————

applicable parameter, rather than weight. It can be argued that the higher the weight the higher the speed, but, in the opinion of the author, this fails to take into account the lift characteristics of the aircraft, and therefore the V_S for V_{AT_0}. It is submitted that, for example, one may consider two separate aircraft types, (for illustrations purposes only) one of which may have a clean, swept wing, and the other a high-lift, high aspect ratio wing. Both have a similar weight for V_{AT_0}. The approach speed of the former

Table 8.2

Cat.	V_{AT}	Initial approach speed range	Final approach speed range	Max. speed for circling	Max. speed missed approach	
					Intermediate	Final
A	91	90/150 (110*)	70/100	100	100	110
B	91/120	120/180 (140*)	85/130	135	130	150
C	121/140	160/240	115/160	180	160	240
D	141/165	185/250	130/185	205	185	265
E	166/210	185/250	155/230	240	230	275

* Max. speed for racetrack procedures. All speeds in knots. $V_{AT} = 1.3 \times V_S$.

will, almost certainly, be higher than that of the other, due to the higher value V_S. Yet, under the weight related categorisation, the slower aircraft must meet the same requirements as the faster. As mentioned earlier, UK requirements directly linked RVR to approach speed, until weight became the controlling parameter (although France had this method of RVR assessment many years earlier). The UK is now phasing out the weight method and is following the ICAO methodology in most significant aspects.

The essence of the ICAO PANS-OPS method is to divide all airliners into five categories, each category covering a range of V_{AT_0} values (1.3 × V_S). Table 8.2 shows how these categories are listed.

Decision Height or Altitude

Some acronyms differ from the CAP 360 method. Decision height (DH) and decision altitude (DA) refer to precision approaches only, while minimum descent height (MDH) and minimum descent altitude (MDA) are used for non-precision approaches and for circling approaches. It must be clearly understood what the difference is between altitude and height. Altitude is the height amsl (QNH) while height is the height above the runway threshold (for precision approaches) or the airport elevation (for non-precision approaches) (QFE). Instead of a value of OCL for arriving at a DH, the ICAO method publishes, for every aid in use at an airport, an OCH (or OCA). This obstacle height (or altitude) is published for all instrument approaches, and includes a margin over the height of the highest obstacle within the defined area for each aid. For precision approaches the margin varies with aircraft approach speed, height loss, altimeter error (or characteristics), glide path angle, and the height of the airport. The obstacle, by the way, may be *either* on the approach or the missed approach, whichever has the greatest effect. Having established the obstacle/OCH (or OCA) margin, thus fixing the OCH/A, the DH/

DA is then derived from the OCH/A, adding a further margin to the latter. This second margin is added to allow for variations in the category of operation, ground and airborne avionics characteristics, crew experience, aircraft performance, and so on. (See PANS-OPS 8168/611, Vol. I, p. 3–5; see also Fig. 8.1.)

Non-Precision Approach
In the case of a non-precision approach there is a basic difference, in that

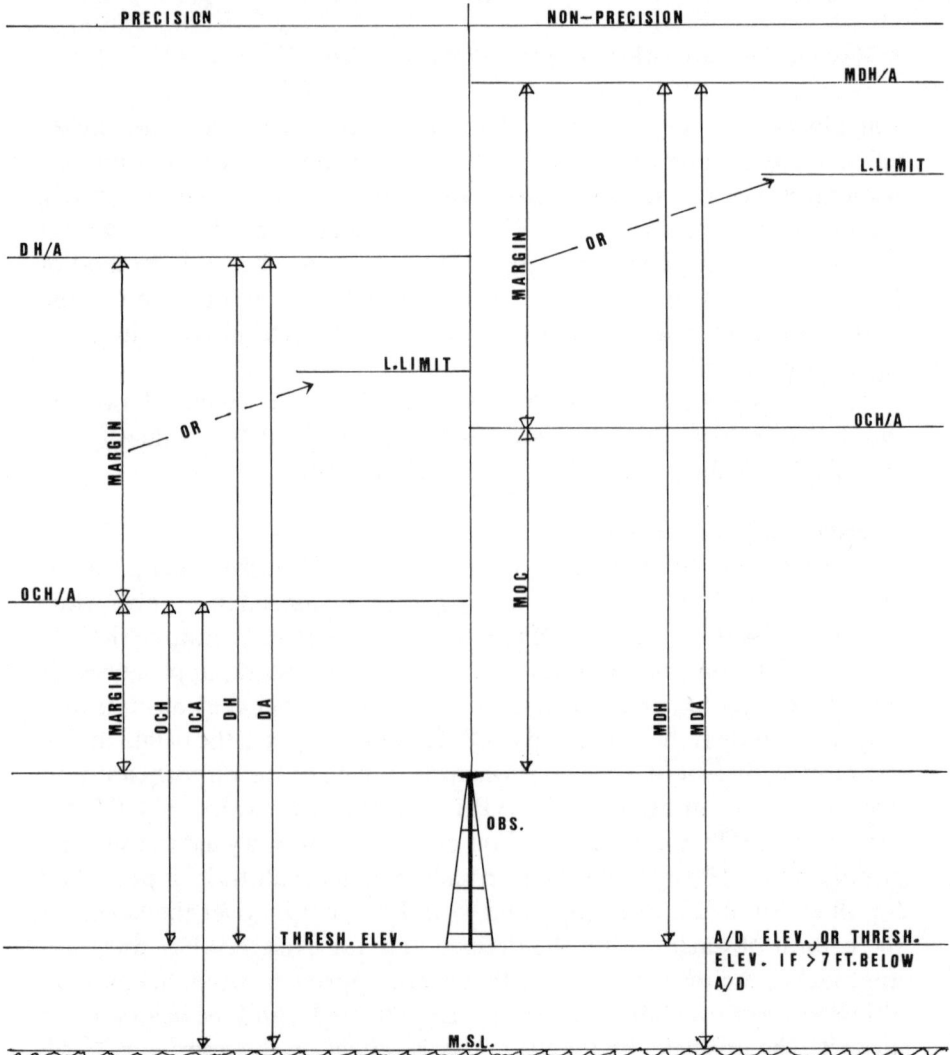

Fig. 8.1 A schematic illustration showing how ICAO arrives at decision height or decision altitudes. See text for details of increments or margins, as appropriate. No particular scale is used in this figure.

the OCH/A is linked to the obstacle height or altitude by a fixed value, known as the minimum obstacle clearance (MOC). The MOC increment varies according to whether or not there is a final approach fix (FAF). That is to say, if the final approach commences from an established fixed position (e.g. from a VOR). Where a FAF is available, the MOC is 246 ft above the obstacle; with no FAF the MOC is 295 ft. (It should be noted that ICAO uses metric values and therefore the rather 'rough' MOCs quoted are the metric values converted to the more normal operational vertical measurements used, namely feet). The height of the significant obstacle plus the MOC gives the OCH/A. The minimum descent height/altitude is the OCH/A plus a margin, as in the case of the precision approach. In both cases – precision or non-precision – the margin between either the DH/DA and OCH/A, or MDH/A and the OCH/A, is not specified by ICAO and is therefore subject to the appropriate state's regulatory authority.

Note
A non-precision approach is normally referenced, as regards the vertical element, to the airport altitude or height. However, if the runway threshold is more than 7 ft below this then the reference point is the threshold (see Fig. 8.1).

Circling
When an approach is made from a circling procedure we find that the vertical elements are far more subject to specification. Firstly, the circling area is defined, according to aircraft category and speed. In somewhat simplified terms, the circling area is contained within the interlinking of the radii from an airport's various runway thresholds. Cat. 1/100 kt gives a radius of 1.68 nm from each threshold, cat. B/135 kt gives 2.66 nm, cat. C/180 kt gives 4.20 nm, cat. D/205 kt, 5.28 nm and cat. E/240 kt gives a radius of 6.94 nm. A wind factor of 25 kt is included when applying the above speeds, which are IAS and at an altitude of 2000 ft amsl. The turn must be made at an average bank angle of 20°, or 3° per second, whichever calls for less bank.

The OCH may not be less than the following heights: cat. A/394 ft, cat. B/492 ft, cat. C/591 ft, cat. D/689 ft, and cat. E/787 ft. The MOC may not be less than: cats A and B/295 ft, cats C and D/394 ft, and cat. E/492 ft. The obstacle is taken from the highest present in the circling area, as defined above. See Fig. 8.2.

Note
In the UK these figures may be rounded-off *upwards* to the nearest 10 ft.

In the UK, at least, the absolute minima for the vertical elements are the

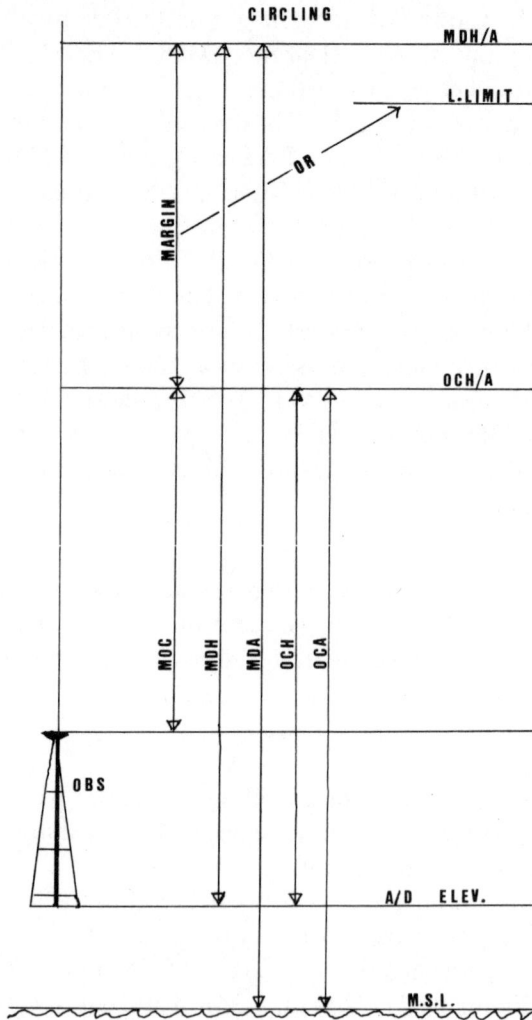

Fig. 8.2 ICAO circling (or visual manoeuvring) min. decision height (or altitude). See text for the vertical values as prescribed by PANS-OPS 8168/611.

same as those listed for the CAP 360 method. Using the ICAO method, unless a margin is specified, each state fixes its own value here and this will either be reflected in that state's regulation of its own aircraft *or* in the derivation, and publication, of state minima.

ICAO PANS-OPS 8168/611 gives no guidance as regards RVR, and this is left – for the time being, anyway – to individual states to decide. In the UK, DH or MDH is applied to Table 8.1 when using ICAO values in exactly the same way as is used when operating under CAP 360 rules. It is understood that ICAO will, in due course, issue guidance on this aspect, but until this happens each and every state will form its own rules. In Fig. 8.3 a table of AOM values for London (Heathrow) is presented, using

SPEED RELATED

(HEATHROW) LONDON
AERODROME OPERATING MINIMA

AO
EFF 25 AUG 88

The following Aerodrome Operating Minima is based on speed related categories. However, to conform to CAP 360 regulations Cat A RVR is given both for A/C not exceeding 5700kgs and exceeding 5700 kgs.
See explanatory notes at front of manual.
<A/C not exceeding 5700 kgs
>A/C exceeding 5700 kgs

Procedure	A				B			C			D		
	DA/ MDA QNH ft	DH/ MDH QFE ft	RVR <5700 kg m	RVR >5700 kg m	DA/ MDA QNH ft	DH MDH QFE ft	RVR >5700 kg m	DA/ MDA QNH ft	DH/ MDH QFE ft	RVR >5700 kg m	DA/ MDA QNH ft	DH MDH QFE ft	RVR >5700 kg m
05 SRA(2nm)	730	650	1100	1500	730	650	1500	730	650	1500	730	650	1500
09L ILS/DME	280	200	600	600	280	200	600	280	200	600	280	200	600
09L LLZ/DME	460	380	750	1000	460	380	1000	460	380	1000	460	380	1000
09L SRA(2nm)	730	650	1100	1500	730	650	1500	730	650	1500	730	650	1500
09R ILS/DME	275	200	600	600	275	200	600	275	200	600	275	200	600
09R LLZ/DME	475	400	750	1000	475	400	1000	475	400	1000	475	400	1000
09R SRA(2nm)	725	650	1100	1500	725	650	1500	725	650	1500	725	650	1500
23 ILS/DME	280	200	600	700	280	200	700	280	200	700	280	200	700
23 LLZ/DME	490	410	750	1300	490	410	1300	490	410	1300	490	410	1300
23 LLZ/DME(1)	640	560	1000	1500	640	560	1500	640	560	1500	640	560	1500
23 SRA(2nm)	730	650	1100	1500	730	650	1500	730	650	1500	730	650	1500
27L ILS/DME	280	200	600	600	280	200	600	280	200	600	280	200	600
27L LLZ/DME	490	410	750	1000	490	410	1000	490	410	1000	490	410	1000
27L SRA(2nm)	730	650	1100	1500	730	650	1500	730	650	1500	730	650	1500
27R ILS/DME	280	200	600	600	280	200	600	280	200	600	280	200	600
27R LLZ/DME	460	380	750	1000	460	380	1000	460	380	1000	460	380	1000
27R SRA(2nm)	730	650	1100	1500	730	650	1500	730	650	1500	730	650	1500

	MDA QNH ft	MDH QFE ft	IFV m	–	MDA QNH ft	MDH QFE ft	IFV m	MDA QNH ft	MDH QFE ft	IFV m	MDA QNH ft	MDH QFE ft	IFV m
All Circling	680	600	1900	–	730	650	2800	880	800	3700	880	800	4600

Notes (1) LLZ/DME without 'NE'.

Fig. 8.3   AOM for London (Heathrow) as presented by AERAD to ICAO specifications. This table includes minima for aircraft of less than 5700 kg in category A. (Courtesy of British Airways AERAD.)

Elev 80	OCH C 2 : A53, B60, C74, D86 C 1 : A144, B154, C167, D183 LLZ 400	(HEATHROW) **LONDON** N Holds I BB 109.5 **ILS/DME 09R**

HEATHROW App/Radar 119.2 127.55 119.5 120.4	Tower 118.7 121.0	Ground 121.9	ATIS 115.1 133.075 113.75	M3 ƎW 25 AUG 88

EGLL

SSA 25nm **22** — W 000° 30' — 20' — 10' — 000° 00' — SSA 25nm **23**

BIG 37d
HENTON 'HEN' 269
142°
122°
232°
322°
052°
302°
BOVINGDON BNN 113·75 Ch 84 (20)

2o Min alt **7000*** 40' at **7000***
BNN 4d
BIG 329R
183°
BOVVA BIG 32d Elstree
CHILTERN 'CHT' 279 at **7000***
181M
222°
11
'OW'
293°
Denham At **4000** 113°
007R
Northolt
Hayes (H)

LAMBOURNE LAM 115·6 Ch 103
16 LON 30d
158°
253°
268° (A)
338°
073°
088°
265° 24
LON 073R
(17)

TAWNY LON 25d Min alt **7000***
1 6 (17)

BNN N15d
51° 30'
LON 11d
'OW' 121M
095° I-BB 7·5d
1 6
BURNHAM BUR 117·1
LON 113·6 Ch 83
'NE' 389·5 L
(18) (17)
London (City)
London Westland (H)
'LCY' 322
(18)
(21)

L 'OW' 334 **ILS D** 'OE' 334 L

(18) D133 1650 Fairoaks
EPSOM 'EPM' 316
BIGGIN BIG 115·1 Ch 98

20 D132 1300
By Notam 2500
D133A 1200
OCKHAM OCK 115·3 Ch 100
164°
274°
344°
094°
BIG 274R
At **3000**
2o

SSA 25nm **21** — — SSA 25nm **23**

I BB 7·5d

2500 2430 ✕
095°
LOM
MM
1320 1250 (LLZ proc)
3°

Ahead to **3000** 2930 then as directed by ATC
Radio Comm Failure
Ahead to LON 10d then right to 'EPM' at **3000** 2930 or below, see chart M11.
When directed to 'CHT' see chart M12.

Var 5°W	GP at MM **340** 270	D THR Elev **75**/3mb	GP at D. THR 52

10 — 5 4 3 2 1 0 1 2 3 4 5

ILS/DME I BB

	T.Lev ATC T.Alt 6000		1. *ATC will allocate appropriate FL. 2. **LLZ proc: **1320** 1250 at LOM gives 3° angle of descent. (State min alt at LOM **1280** 1210). MAP at MM. 3. DME frequency paired with ILS and indicates zero at threshold. 4. If LAM u/s hold on TAWNY. 5. If BNN u/s hold on BOVVA. 6. Max holding speed IAS 220kt, except at Lambourne Max IAS 240kt.		

kt	fpm	LOM		
200	1060	-	7.5d	**2500** 2430
180	950	DTHR	7d	**2340** 2270
160	850	1:25	6d	**2020** 1950
140	**740**	**1:37**	5d	**1700** 1630
120	640	1:54	4d	**1380** 1310
100	530	2:16	3d	**1060** 990
80	420	2:51	2d	**740** 670
			1d	**420** 350

Rev: Editorial

© BRITISH AIRWAYS AERAD

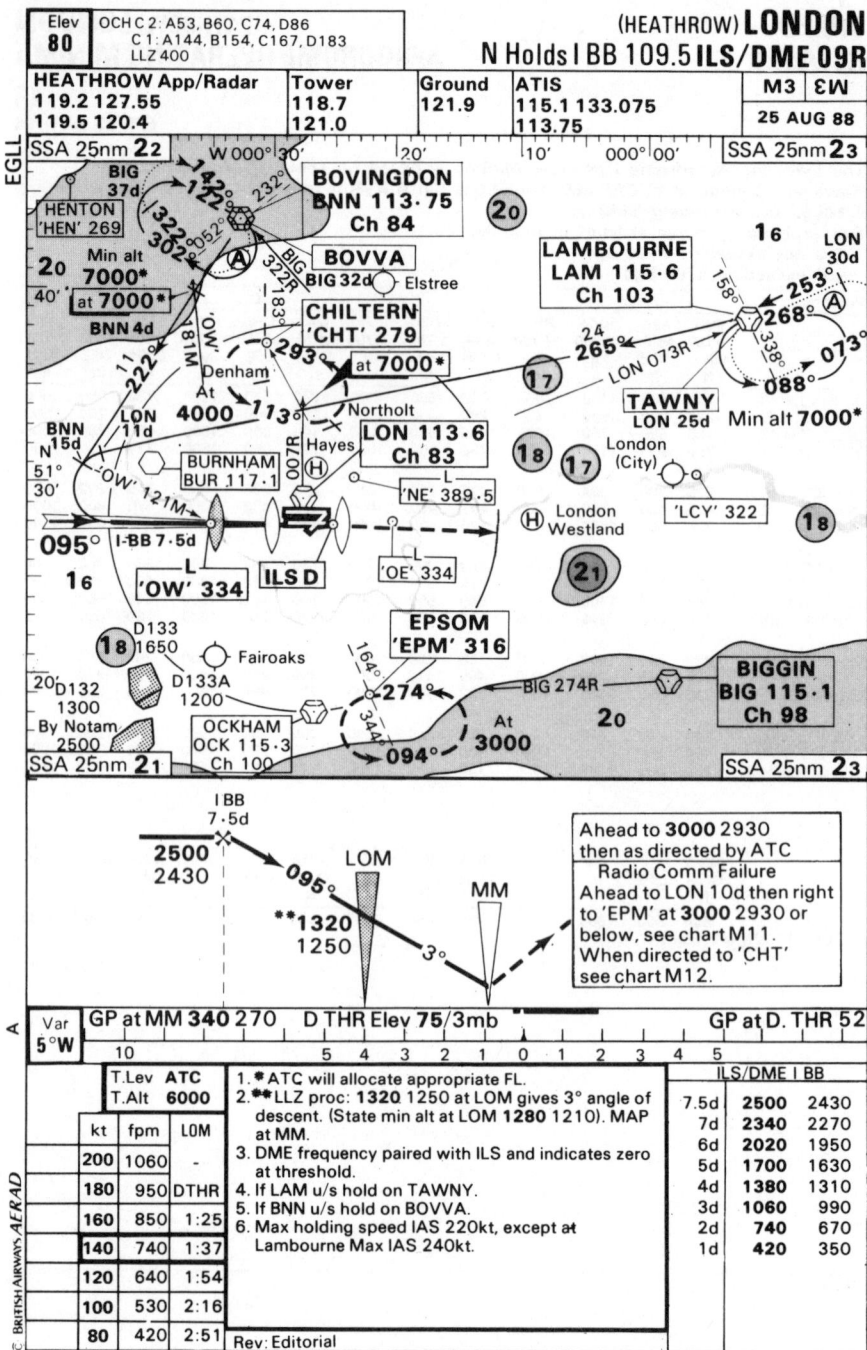

Fig. 8.4 A typical instrument approach procedure chart, as published by AERAD. In the top left-hand corner are boxes containing the airport elevation in feet (80) and OCH values for category 1 and 2 conditions, covering aircraft categories A to D. (Courtesy of British Airways AERAD.)

DAN-AIR SERVICES LIMITED

AIRFIELD OPERATING MINIMA

AIRCRAFT TYPE: BAe 146

AMENDMENT NO.   169 - 09.11.88

COUNTRY: SWITZERLAND

AERODROME: BERNE

ELEVATION: 1673 FEET

AIRFIELD CATEGORY: "C"

PAGE 1 OF 3

RUNWAY LENGTH METRES	APP. LIGHTS	AID	TAKE-OFF		LANDING		VISUAL MANOEUVRING		NOTES
			CC FEET	RVR METRES	DH FEET	RVR METRES	DH FEET	IFV METRES	
14 (1310)	HI	ILS/DME ILS/DME NDB	NIL	350	390 720 810 970 1130	1500 1500 1500 1500 2850			NOTE 1 NO 'GP' NOTE 2 NOTE 3
32 (1310)	NIL		NIL	1200	1180	1500			CIRCLING APPROACH
CIRCLING							1180	2800	CIRCLING RADIUS 3 NMLS TO THE NORTH EAST OF AIRFIELD ONLY.
			STATE MINIMA						

AIRFIELD CLEARANCE - SPECIAL BRIEFING - ISSUE NO: 10/NOV 88/CAF

The airfield is in a bowl surrounded by high ground and located 5 N.Mls. South East of the City and to the South and West of the River Aare. High ground rises to above 3000 ft. AMSL within 8 N.Mls. to the East, 4 N.Mls. to the South and 3 N.Mls. to the West. A ridge rises to 2770 ft. AMSL within 3/4 N.Mls. West of and parallel to the R/W 14 extended centre line approx. 1 3/4 N.Mls. from the R/W 32 Threshold.
Note the safety altitude of 7.2 approaching WILLISAU, the BERNE 25 N.Mls. Sector Safe Altitudes, and the maximum speeds in the "SHU" and "BER" holding patterns.

SPECIAL RESTRICTIONS: (1) NIGHT LANDINGS PROHIBITED.
                        (2) CAPTAIN - AIRFIELD CHECK AS OPERATING PILOT.
                        (3) MAXIMUM CROSSWIND PERMITTED FOR LANDING - 25 KNOTS.
                        (4) MINIMUM BRAKING ACTION MEDIUM/GOOD.

NOTE 1:   (a)   Until commander has completed 4 ILS/DME approaches to R/W 14 ADD 250 ft. ie. DH.640

Note 2:   (a)   Twin ADF required
          (b)   Commander completed 2 NDB approaches at BERNE in last 6 months.
          (c)   Aerodrome Lighting System (ALS) and PAPI must be operative. Request threshold lights.

Note 3:   (a)   Twin ADF required
          (b)   Commander completed 2 NDB approaches at BERNE in last 12 months.
          (c)   Request ALS, PAPI and threshold lights.

AIRCRAFT PROCEDURE - APPROACH: When requested and approved by Berne ATC, it is permitted to fly the 11 nmls DME arc to intercept the inbound track. Select Flaps 18° before reaching the BIRKI INT (or "SHU" NDB) inbound. Select gear down after passing BIRKI INT (or "SHU" NDB) inbound and Flap 24° intercepting the glide scope, or before reaching "MUR" NDB if "NO G.P." as NDB letdown. This procedure will allow a more controlled approach for either runway.
For landing on R/W 14 select Flap 30° passing 1000 ft and Flap 33° after passing "MUR" NDB.
For landing on R/W 32 select Flap 30° on commencing the turn and the descent. Select Flap 33° when the final approach has been assessed.

CONTINUED /....

Fig. 8.5   An alternative way of presenting AOM, in this case by means of the route book. An advantage of this method is that it permits the presentation of special briefing, when required, in the same table. (Courtesy of Dan-Air Services Ltd.)

DAN-AIR SERVICES LIMITED

AIRFIELD OPERATING MINIMA   AERODROME: BERNE

AIRCRAFT TYPE: BAe 146

AMENDMENT NO.   168 - 28.10.88   PAGE 2 OF 3

CONTINUATION OF AIRFIELD CLEARANCE - SPECIAL BRIEFING - ISSUE NO.10/NOV 88/CAF

AIRCRAFT PROCEDURE - TAKE-OFF: All take-offs must be carried out at rated N1 Power from a standing start. Note take-off minima R/W 32 due to proximity of glidescope aerial.
There is an emergency turn procedure on R/W 14 and R/W 32.

INSTRUMENT APPROACH - ILS/DME R/W 14: Descend in BIRKI hold to 4000 ft. After passing BIRKI inbound 7.4D intercept glidescope 5.4D and check height 1320 ft. QFE passing 3.0D, ABM "MUR" NDB at 880 ft. QFE, to company minima (see NOTE 1).
If visual at D.H. note - WARNING - Do not undershoot PAPI due to OBST after D.H.
If no visual contact, climb ahead to abeam "BER" NDB (IBE 0.8D) and proceed as the missed approach procedure.

INSTRUMENT APPROACH - LLZ/DME (No G.P.) R/W 14: After BIRKI inbound, descend to conform to a 4.0° descent, crossing 3.0D 1320 ft 2.0D ABM "MUR" NDB 880 ft to minima of 670 ft.

INSTRUMENT APPROACH - NDB R/W 14: Descend in "SHU" hold to 4000ft. after passing "SHU" descend on track to reach decision height of 1130 ft, 970 ft. or 810 ft. at or just before the "MUR" NDB (see NOTES 2 & 3). Advise APP before commencing if using D.H. below 1130 ft. The "MUR" is 500 metres to the left of the R/W centre line and 3500 metres from the R/W 14 threshold. After passing the "MUR" it will be necessary to turn right to align on the R/W 14 centre line (141°M) and follow the PAPI (4.0°). There is a flashing light on each side of the threshold. If no visual contact over "MUR" NDB, climb to the "BER" NDB and proceed as the missed approach procedure. Maximum speed 160 knots, until left turn completed, and maximum climb gradient until reaching 4000 ft.

VISUAL APAPROACH - R/W 14: If visual with the R/W before reaching the "MUR" NDB, proceed on the R/W centre-line but note the prominent cathedral spire 2100 ft. AMSL just to right of the centre-line, and tall trees to the left of the R/W centreline do NOT descend below the 4.0° slope of the PAPI.

VISUAL APPROACH - R/W 32: Proceed as instrument approach to "MUR" at 1200 ft. QFE. From "MUR" track 125° until 220° QDM "BER" NDB and commence turn onto finals. There is no approach lighting, but there are VASIS (3.25°) on each side of the runway which are aligned for the displaced threshold of 110 metres. The full length of R/W is available as the barrier on the road will be down for the approach. NOTE: Runway width is only 30 metres (98 feet).

MISSED APPROACH - R/W 32: If visual contact with the runway is lost, continue towards the airfield as above, but commence climb to 4000 feet. Maximum speed 160 knots until right turn is completed. Continue tracking along the runway centreline to intercept 328° QDM "SHU" NDB, to "SHU" NDB, or to back track along the LLZ to BIRKI.

ATC: Control very good. There is intensive glider and light aircraft activity particularly to the North and West of the airfield especially at weekends.

ATIS: 125.12

CONTINUED/...

Fig 8.5   Continued

DAN-AIR SERVICES LIMITED

AIRFIELD OPERATING MINIMA   AERODROME: BERNE

AIRCRAFT TYPE: BAe 146

AMENDMENT NO. 168 - 28.10.88   PAGE 3 OF 3

CONTINUATION OF AIRFIELD CLEARANCE - SPECIAL BRIEFING - ISSUE NO10/NOV 88/CAF

WEATHER: The weather is generally good. The main problem being stratus during the winter months with a base about 1000' AAL. This occurs with a low over Spain and a high over Austria and a Southerly wind in the local area. Thunderstorms occur occasionally, with gusting crosswinds during the Summer, however, these transit quite rapidly with clearance in about 20/30 minutes. Winds are 90% below 9 knots, 6% 10-15 knots with the predominant directions 240° through North to 040° and 140° to 180°.
Temperatures can rise to an average of +25°C in the afternoon in July and August, with a possibility of the temperature reaching +29°C on at least 7 days of the month. It may reach +36°C under extreme conditions.

GENERAL: Caution when taxying onto the apron due to light aircraft. If possible proceed clockwise to park outside the terminal. Do not use the yellow taxiway centreline (parallel taxiway) which is for GA aircraft only.

GROUND HANDLING: By CROSSAIR good. (135.75)
If weather prevents operation into BERNE, the decision must be made before departure from LGW that the destination becomes BASLE or ZURICH. CROSSAIR will then arrange for passengers to be taken to BASLE or ZURICH to minimise delays.

Fig 8.5 Continued

PANS-OPS methodology for the vertical elements (DH, MDH etc.) and applying the values so obtained to the CAP 360 Table 8.1. A specimen instrument approach procedure using the UK formula is shown at Fig. 8.4. Here we have a full ILS, backed up by DME, as the approach aid. In the top left-hand corner is a box that gives the airport elevation amsl, namely 80 ft. Next to it is a further box that gives the AOM Cat. 1 values for the various speed-related aircraft categories, thus: cat. A – OCH 144 ft, cat. B – OCH 154 ft, cat. C – OCH 167 ft and cat. D – 183 ft. With LLZ only (i.e. no G/P) the OCH for all categories is 400 ft.

Note
Both Figs 8.3 and 8.4 are published by British Airways AERAD who have decided only to publish for cat. E to special order, due to the rarity of the number of aircraft that fall into this category, e.g. Concorde or the Lockheed Galaxy. It should also be noted, when considering Fig. 8.3, that there is no significance in the heading of columns with, say, DH/MDH and then giving RVR, when a non-precision aid is involved. This is purely an editorial method of layout in a single tabular format, and DH/RVR (or DA/RVR) applies only to *precision* aids, while MDH/RVR, MDA/RVR etc. applies to non-precision aids. Figure 8.5 shows an alternative way of presenting AOM for a *specific aircraft*.

PANS-OPS does not specify, as yet, takeoff minima, and therefore individual states deal with this in their own way. Using PANS-OPS methodology the UK takes a similar attitude as it does when using CAP 360, as detailed earlier.

Categories 2 and 3
So far we have dealt exclusively with Category 1 AOM, that is, when the DH minimum is 200 ft and the associated RVR is not less than 600 m. This is because Cat. 1 is not only the most widely used set of meteorological conditions but also it is necessary to fully understand Cat. 1 before attempting to examine Cat. 2 and Cat. 3. Cat. 2 allows for DH values below 200 ft, down to 100 ft, while Cat. 3 permits a DH of less than 100 ft, or even no DH. Cat. 3 is sub-divided into three subcategories, viz: Cat. 3a – DH less than 100 ft, down to 50 ft; Cat. 3b – DH less than 50 ft, and Cat. 3c – no DH (and with which can be associated 'RVR' – the so-called 'zero/zero' category). No specific values of RVR are prescribed, but the guidance is given as to suitable values. For Cat. 3a the minimum should, at the DH, provide enough visibility so that, in the event of ILS failure, a safe landing can be carried out. *It must be clearly understood that both Cats. 2 and 3 rely on automatic approaches and landings, using suitably accurate and reliable autopilots and both airborne and ground systems.* (Single installations are not acceptable.)

For Cat. 3b the appropriate RVR values lie within the range 150–200 m, but may be less if the aircraft is equipped with suitable ground roll guidance means. As regards Cat. 3c there is no DH, and consequently no RVR requirement. The aircraft must be equipped with a 'fail-operational' ground roll guidance system. We are now talking about landing totally automatically, including auto-throttles, radio altimeters, ground roll guidance, and anti-skid brakes. The cloud ceiling can be at runway level, while the visibility on the runway can be nil. The main problem with Cat. 3c is not landing but taxying in afterwards! Bear in mind that, in zero visibility, even airport vehicles cannot see. Airport surface movement indicator radar can help, but even then only to a limited extent.

It is recommended that keen students of AOM should read the following documents: ICAO PANS-OPS 8168/611 (and, in particular, Vol. I, part 7 which, at the time of writing, has still to be developed. This will fully define ICAO AOM, when published). Also of interest is JAR–AWO (Joint Airworthiness Requirements – all weather operations), the European code of airworthiness. See also Fig. 8.6.

Fig. 8.6   Where ATC are almost in IMC, but the aircraft is not. A BAe 146 of Dan-Air at Amsterdam (Schiphol). The airport is 13 ft below amsl. (Courtesy of Dan-Air Services Ltd.)

Table 8.3 Example of statistical observations covering values of visibility and cloud base. (Courtesy of Austrian Meteorological Office.)

Model A

Station oder Flughafen ... SALZBURG	
Station or aerodrome	
Monat ... FEBRUAR	Beobachtungstermin ... 1800 ...GMT
Month ... FEBRUARY	Time of observation ...GMT
Geogr.Breite Latitude ... 47° 48′ N	Geogr. Länge Longitude ... 13° 00′ E	Beobachtungsperiode Period of record ... 1955–1971
Seehöhe Elevation ... N.N above MSL. ... 430 m	Gesamtzahl der Beobachtungen Total number of observations ...

Mittlere Anzahl der Tage pro Monat von gleichzeitigem Auftreten bestimmter Sichtbereiche und bestimmter Bereiche der Untergrenze der tiefsten Wolkenschicht, welche mehr als 4/8 des Himmels bedeckt.
Mean number of simultaneous occurrences of specified visibility ranges and specified ranges of the height of the base of the lowest cloud layer covering more than 4/8ths of the sky.

h_s h_s W	0–30	30–60	60–90	90–120	120–150	150–180	180–240	240–300	300–450	450–900	900–2400	≥ 2400	$n_s < 5/8$	Total
0–0.1														
0.1–0.2														
0.2–0.3			0,1										0,1	0,2
0.3–0.4													0,1	0,1
0.4–0.5				0,1		0,1			0,1				0,1	0,4

h_s / W	0–30	30–60	60–90	90–120	120–150	150–180	180–240	240–300	300–450	450–900	900–2400	≥ 2400	$n_s < 5/8$	Total
0.5–0.6														
0.6–0.7							0,1	0,1					0,1	0,3
0.7–0.8			0,1										0,1	0,2
0.8–0.9								0,1					0,2	0,3
0.9–1.0														
1.0–1.2		0,1					0,1	0,1	0,1				0,1	0,5
1.2–1.6		0,1					0,1	0,2	0,1	0,1	0,1	0,1	0,1	0,9
1.6–2.0						0,1						0,1	0,7	0,9
2.0–2.4							0,1	0,2	0,2	0,1	0,1	0,1	0,4	1,2
2.4–3.2							0,1		0,3	0,4	0,2	0,2	0,6	1,8
3.2–4.8									0,2	0,7	0,5	0,6	0,8	2,8
4.8–8.0									0,2	0,9	2,1	0,7	1,8	5,7
8.0–16									0,1	1,1	3,5	1,4	2,1	8,2
≥ 16											0,8	1,6	2,3	4,7
Total		0,2	0,2	0,1		0,2	0,5	0,7	1,3	3,3	7,3	4,8	9,6	28,2

Statistical Meteorological Data

Reference has already been made in this book to meteorological statistics, mainly in connection with winds and temperatures. The tabulation of specified meteorological data is made in a specified format to the World Meteorological Organisation (WMO). In Table 8.3 an example is given of a page of statistical observations covering values of visibility and cloud base, averaged for the month of February. The time of these observations is 1800 GMT, and the period is from 1955 to 1971. The example shows the conditions returned for the appropriate period by Salzburg, Austria.

9: Emergency Procedures

Caution
This chapter is provided for illustration purposes only and must not be used operationally as it is incomplete.

Emergencies are occurrences that have a habit of cropping up in all walks of life, and they have no respect for either rank or position. They may be as a direct result of a malicious act, such as sabotage, or as an Act of God, or simply directly attributable to incompetence or lack of self-control. They can be such that only when they occur does it become evident that precautionary procedures were lacking something and rectification action is then immediately taken – action that should have been taken before the emergency. Sometimes an emergency may cause acute embarrassment to a few individuals, and nothing more serious, while to the majority it is seen as a source of mirth. In a well-run airline it pays (and is morally vital) to anticipate contingencies, and to define certain circumstances that must be construed as being emergencies, together with clear and unambiguous instructions as to the action to be taken should one or more of these emergencies actually take place. Sadly, the past is littered with records of aircraft accidents that need never have taken place, often caused by human error or lethargy. Yet, in some cases, the competent authority has made clear, and strong, recommendations after an accident, only to have these totally ignored by another authority. Often inter-departmental pride is to blame in such cases, and this is inexcusable. All should pull together to avoid a similar accident ever taking place again.

A flight manual will normally contain a section devoted to most foreseeable emergencies that *could* occur to that aircraft at some time or another. It will clearly define each emergency, together with the action to be taken, either to contain the emergency or to minimise its effect. One problem that can manifest itself is the heavy, official type of language used in this document, and this needs some simplification into ordinary everyday language. The other consideration is that, should an emergency arise, it is far preferable *on all counts* that the appropriate information regarding the action to be taken should immediately be to hand. It is *not* a good practice to have to seek a different book from the library (on board, of course) in order to check on the actions to be taken, according to the nature of the emergency. Thus, emergency procedures need to be translated into succinct terms and to be made immediately available through the means of a document that is in fairly continuous use, i.e. the route book. In the pages that follow an example will be given of how an emergencies section of the route book could be presented, albeit in condensed form, and not necessarily complete. The subject aircraft is, once more, the BAe 146-100, for the sake of continuity.

The layout for an emergencies section of the route book can look something like this, as in the case of limitations (see Chapter 5).

BAe 146-100 (ALF-502R-3 engines)
(Certain customer options may need to be included, according to individual aircraft.)

1. Emergency Landing

1.1 Emergency Landing – Initial Action
(a) ATC transponder -- set to A 7700.
(b) Distress signals – transmit. Give position, nature of emergency, and intention. Give estimated time to landing. (*Use 121.5 MHz if not in contact.*)
(c) Brief cabin staff. Give nature of emergency, intention, and estimated time to landing.
(d) Loose equipment – secure.

1.2 Emergency Landing – Descent
(a) Pressurisation – reset.
(b) Flight instruments – set and crosscheck.
(c) Special procedures – check safety height and approach.
(d) Landing – check data and set bugs.
(e) Warning signs – no smoking and fasten seat belts *on*.
(f) Ground proximity warning – CBs A and B21 *pull*.
(g) Seats and harnesses – checklocked and secured.

1.3 Emergency Landing – Approach
(a) Radar – *off*.
(b) Altimeters – set and crosscheck.
(c) Landing lights – *on*.
(d) Flight deck emergency lights – *on*.
(e) Brake fans – *on*.
(f) AC pump – *on*.
(g) Fuel state – check.
(h) APU air – *off*.

(i)   APU – *stop.*
(j)   Cabin emergency lights – *on.*
(k)   Beacon – *off.*
(l)   Strobe – *off.*
(m) Nav lights – *off.*

Note to 1.3 (k) to (m)
Only when landing on unpaved surface, to minimise fire risk. Leave *on* if landing on paved surface, unless risk of U/C collapse.

(n)   Slides – check *arm.*
(o)   2000 ft – gear *down* and checked (*excluding* ditching).
(p)   Brakes – check.
(q)   Altimeters – check *set for landing.*
(r)   Engine air – *off.*
(s)   Pressurisation – *manual – open.*
(t)   Flaps – landing.
(u)   Shoulder harness – locked.
(v)   1000 ft – PA announcement 'Take up emergency positions'.
(w) 200 ft – fuel pumps *off.* PA announcement 'Brace, brace'.
(x)   After landing roll – parking brake – *on.*
(y)   Thrust levers – *fuel off.*
(z)   Batteries – off when no longer required.

Note
Fire handles should be pulled (1,2,3 and 4) and rotated to Ext 1 or Ext 2 if required.
  Try to end landing with nose into wind, if possible, to reduce fire risk.

1.4 Evacuation
(a)   Assist cabin crew to evacuate pax.

1.5 Emergency Evacuation
(a)   Thrust levers – *off* as soon as possible.
(b)   Cabin crew – notify, and check Slide arming handles to *arm.*
(c)   ATC – notify.
(d)   Wheel brakes handle – *park.*
(e)   Pressurisation – *manual-open.*
(f)   Flight deck emergency lights – *on.*
(g)   Cabin emergency lights – *on.*
(h)   Flight deck window(s) – *open.*
(i)   Fire handles – pull to full extent. Turn to Ext 1 and Ext 2 if required.

(j) APU – *stop*.

(k) APU fire extinguisher – *disch* (if required).

(l) Cabin crew – commence evacuation procedures.

(m) Batteries – *off*.

2. Ditching
(Emergency landing on water)

2.1 Ditching – Initial Action
(a) ATC transporter – set to A 7700.

(b) Distress signals – transmit. Give position, nature of emergency, and intention. (Use 121.5 MHz if not in contact.)

(c) Cabin crew – advise nature of emergency and intentions, with estimated time to touchdown.

(d) Loose equipment – secure.

(e) Emergency equipment – prepare.

2.2 Ditching – Descent
(a) Life jackets – *on*.

(b) Pressurisation – reset.

(c) Flight instruments – set and crosscheck.

(d) Briefing – special procedures including safety height and approach.

(e) Landing – check data and set bugs.

(f) Warning signs – no smoking and fasten seat belts – *on*.

(g) Aural warning CBs (A and B29) – *pull*.

(h) Ground proximity warning (CBs A and B21) – *pull*.

(i) Seats and harnesses – check locked and secured.

2.3 Ditching – Approach
(a) Check sea state from 500 ft to 1000 ft. Swell direction.
Wind direction from direction of waves.
Wind speed from sea state and crests.

(b) Prepare to alight *along* swell; *not* across. Preferably on crest or back.

(c) Radar – *off*.

(d) Altimeters – set QNH and crosscheck.

(e) Landing lights – *on*.

(f) Flight deck emergency lights – *on*.

(g) Fuel – check state.

(h) APU air – *off*.

(i) APU – *stop*.

(j) Cabin emergency lights – *on*.

(k) Slides – as required.

Note

If slide rafts carried check slides *armed*.

If separate life rafts carried, slides *disarmed*.

(l) 2000 ft – check gear *up*.

(m) Engine air – *off*.

(n) Pressurisation – *dump/ditch* (for zero pressure differential).

(o) Flaps – 33°.

(p) Shoulder harnesses – *lock*.

(q) 1000 ft – PA announcement 'Take up emergency positions'.

(r) 200 ft – fuel pumps – all *off*.

(s) PA announcement – 'Brace, brace'.

(t) Alighting – min. rate of descent, 12° nose up pitch, if possible.

(u) At rest – thrust levers – *fuel off*.

(v) Fire handles 1,2,3,4 – *pull* (fullest extent). Rotate as required to Ext 1 or Ext 2.

(w) Batteries – *off*.

(x) Assist cabin crew to evacuate pax.

Note

Use of front two doors for evacuation will maximise flotation time, but use doors as dictated by cabin conditions, sea state, and hull attitude.

3. Loss of Cabin Pressure

3.1 *Initial actions*

(a) Crew oxygen masks – *don*.

(b) Crew communications – *check*.

(c) Pressurisation – mode selection – *man. Man shut/open – shut*.

(d) Cabin altitude – *check*.

(e) Cabin rate of climb – *check*.

(f) Warning signs: no smoking and fasten seat belts – *on*.

(g) Pax oxygen drop out – 14 000 ft cabin alt.

(h) Consider emergency descent.

3.2 *Emergency Descent*

(a) PA announcement – emergency descent.

(b) Thrust levers – flight idle.

(c) TMS DISC button – press.

(d) Airspeed – MMO/VMO.

(e)   Airbrakes – *out*.

(f)   Min. safe altitude – check.

(g)   *Cont. Ign* A and B – *on*.

(h)   ATC transponder – set to A7700.

(i)   Rough air speed – Max. 240 kt or M 0.60.

4. Loss of All Generators

4.1 Initial Actions

(a)   Standby generator – check functioning.

(b)   Standby generator functioning – all other than essential services – *off*.

(c)   If standby generator inop, select – *off*.

(d)   Select essential electrical services thus.

Eng 3 – no restrictions on N2

Left pitot heater – *batt* if operationally necessary

Cabin emergency lights – *on* (night only)

Flight deck emergency lights – *on* (night only)

APU – do *not* start

PTU – *off*

Start master – *on* for 5 secs, then *off*

AC *bus tie* – *open*

Gen 1 – *off/reset*, then *on*

Gen 4 – *off/reset*, then *on*

APU *Gen*, if already running – *off/reset*, then *on*.

(e)   If no AC busbars recovered:

APU (if running) – *stop*

Fuel quantity switch – use as required.

4.2 Approach and Landing

(a)   GA *power* – bug as TGT (see GA *power* emergency level).

(b)   Flight idle baulk – at 60% N2 in flight.

(c)   Flaps – remove baulk manually; check flap selections manually.

(d)   Gear – refer to abnormal procedures section (*not included in this book*).

(e)   Brakes – *yel*.

(f)   Anti skid – *batt*.

(g)   Pressurisation – select *man-open* at low p.

In concluding this chapter we must return to its beginning, by inserting a note of caution. Most of the text herein has been taken more or less verbatim from the BAe 146 operations manual, as issued by BAe (whose assistance in this matter the author gratefully acknowledges). However, Chapter 9 is intended as an illustration only, and some further simplification has taken place. This has mainly been done because certain drills laid down by the BAe operations manual emergencies section assume some practical knowledge of the aircraft, and to include these in toto could possibly lead to some confusion. There have been deliberate omissions, therefore, and Chapter 9 must *not* be taken to be an emergencies section for a BAe 146 route book.

In the BAe 146 operations manual there is also a section that deals with abnormal procedures, and this too has been omitted. While we are using the BAe 146–100 as an example aircraft for the purposes of this book we are not purporting to present any guide to the operation of this aircraft, and *this book must not be used in any way whatsoever for such purposes.* Rather, it presents advice that is broadly applicable to almost any aircraft in the airliner category, and is relevant to the subject of route (or operations) planning, using the BAe 146 *as an example only.*

10:   Route Monitoring

Having introduced an aircraft on to a particular route it might be thought that all that now remains is for the route to be operated in accordance with the procedures and data contained in the second part of this book. While this is strictly true, it must be borne in mind that physical changes may well take place as time passes that change the actual physical conditions of the route, and thereby cause the data contained in the route book to become obsolete, or out of date, at least in places. Thus, a very strict eye must be kept on the route characteristics, and Notams should be scanned frequently (daily is not too frequently) for any factors that can affect the route book data. For example, a runway length may change (due to, say, repair or lengthening work) and may become less than that published by the airline. It is clear that to ignore such a situation could be potentially dangerous. New danger areas may be promulgated by a state, and even something quite domestic, such as an airport running out of fuel, has been known to happen. The author remembers only too well arriving at a certain airport for a refuelling stop only to find that, due to weather conditions, fuel deliveries could not be maintained and the airport had virtually no fuel. A diversion was made, with adequate fuel to accomplish this safely, but the diversion airport had, for some reason, closed down its sole instrument descent and cloud break guide. Using DR navigation and mapping radar a cautious descent through cloud was made, but no airport was in sight. Neither was there any radio contact. A large river had been located by radar and after some searching at relatively low altitude, the airport was found and a landing made – with less than 20 minutes fuel remaining. Fortunately this was a ferry flight. A Notam had been sent out regarding the fuel situation at the destination airport, but this had not been received prior to departure.

Another aspect that well repays monitoring is manufacturer's data relating to the aircraft. This is often evolved as a direct result of feedback from operators. A typical example of this could be a change in the recommended engine settings, for commercial reasons affecting the engine overhaul costs or time between overhaul (TBO). But, what is sauce for the goose may not necessarily be sauce for the gander. That is to say, the revised engine settings may not necessarily be beneficial to all operators, depending on average route temperature conditions, fuel prices, or whatever. It therefore can be a cost-effective exercise for route planning to carry out on-line checks to ascertain the effects of the proposed new engine settings. Unless safety is a factor, e.g. the measures are promulgated to prevent occurrences such as a turbine-blade shedding – it may be found that the previous settings give better results for certain operators, as described above. Or it may be worthwhile carrying out experiments (always within the engine handling limitations) to find out an optimum setting for a particular route or routes. Naturally, route planning should carry out such experiments in close liaison with the engine manufacturers. Questions must be asked of the appropriate departments within the airline, such as 'Is route time saving at the cost of increased burnoff preferable, and, if so, to what extent?' Or, 'Is an increased TBO for the engines worth an increased block time, or is the reverse the better situation?' These are but two illustrations to show that route planning's task should not be passive and restricted to the preparation of data or commercial quotations.

Another aspect worth checking is the route profile in practice. What is the optimum FL between two airports, as a function of distance and, perhaps, OAT? If a good lookout is kept on traffic flows across a given FIR or series of FIRs it may be possible to discover if there is a particular band of FLs that permits the regular clearance of 'direct to' routings. That is to say, to take a direct track from a particular navaid to another, by-passing the more normal routing. In the event of such 'direct to' clearances resulting in reduced flight times it may be worth planning for the appropriate FLs as a matter of course. That is to say, file the flight plan for the FL at which the issue of a useful 'direct to' clearance has been established to a reasonable degree of regularity. In the event that such a clearance is not forthcoming en route a request can always be made for a change to a more acceptable FL for the route to be followed.

Another aspect of route monitoring that can repay closer attention with high dividends, is that in which the runway lengths are either limiting or could become so. For example, although we have already discussed temporary runway limitations due to such things as repairs and so forth, sometimes the normal runway lengths can be exerting a limit that need not continue to be so, if a physical on-site check is made. Or, perhaps, the aircraft manufacturer may announce an approved increase in MTWA or MLWA, and an operator may not be able to take any, or full, advantage of this due to runway lengths. In either case a closer look at the runway(s) that are affected is probably justified. All too often, the airport or state authority, promulgates runway data based on the balanced field concept (ASDA = TODA). Do not accept such promulgated data without question, although there is little that an operator can do as regards the *engineered* element of any runway. But sometimes it is possible to liaise with an authority so that clearway may be either published or extended (it is suggested that the reader refers back to Chapter 6 at this stage, to refresh his, or her, memory).

A procedure that the author has found to be practical and rewarding, is to first of all identify limiting, or potentially limiting, runways – that is, runways that already *are* limiting or will not allow for any use to be made of any MTWA that may be authorised. A preliminary study of the obstacle charts in the appropriate AIP may give an initial indication as to the feasibility or otherwise of taking the matter further. Or reference can be made to a suitably-scaled topographical chart that covers the climb-out end of the subject runway, and beyond. Look for evidence of obstacles of a *permanent* nature, such as hills, buildings, and so forth. If there is a hill, or rising ground, or buildings, within the defined clearway area, and the height of such obstacle above the end of the TORA penetrates the required clearway plane surface of 1.25%, then the outlook is not very promising – unless the airport authority can, and will, eliminate or reduce the obstacle so that the clearway plane is not penetrated any longer. If the operator is substantial and is a regular user of the airport, then there is still a chance. Sometimes in the area that appears to be suitable for the declaration of clearway, any obstacles present may be in the form of vegetation, e.g. trees or bushes, and the chances are that these will not be identifiable from any charts. This is where route monitoring can play an active part. A personal visit by a technical pilot, or route planner, can frequently achieve results that cannot be obtained by correspondence. It is good policy, in such cases, for the airport authority to be advised of such a proposed visit and its purpose. It is also necessary for the appropriate 'tools' to be available, either at the airport in question or taken on the visit. The main item required is an instrument for measuring the angle between the extended runway centre line, and the lateral limits of the proposed clearway area, as defined by regulation, to any obstacles lying within this area. From the end of the TORA and sides as defined, identify any obstacles that penetrate the 1.25% plane; the clearway area definition is contained in ICAO Annexe 14 and broadly speaking it is measured from the corners of the cleared 'strip' containing the runway, diverging from the parallel sides outwards by 15° on either side.

A suitable instrument for this purpose is a hand-held sextant or a surveyor's level (e.g. a 'dumpy level'). Using either instrument (or any suitable instrument) scan the proposed clearway area (which will be roughly conical in shape, as shown above) and note any obstacles that are significant. If none exist then the technical job is done, and the declaration of clearway must be negotiated with the airport authority, invoking the authority's assistance and co-operation. The object now is to (a) get the authority to promulgate the clearway, either as such or as an increase, and

(b) to get this in writing, so that immediate use may be made of the new dimensions. Remember, clearway may not exceed 50% of the TORA. However, if obstacles are found to exist within the defined area, the outcome must largely lie in their physical character. A tree or bush can be removed or cropped, so that it lies well below the clearway 1.25% plane slope, and the airport will probably agree to do this, especially if the operator agrees to meet at least part of the cost. But, if the obstacle is, say, a tall chimney, this will certainly involved third parties, and the only hope is for the airport to persuade the owner thereof to either shorten his chimney or to remove it altogether. Here a lot will depend upon the powers of the authority, and type of state government.

The object of the exercise is to permit the airline operator to have his aircraft taking off at a higher than hitherto permitted weight. The economics of this will dictate the extent and vigour of the operator's efforts with the airport authority. The author has had useful successes in the past using this type of approach, and it must be said that nowadays most airports of any consequence are scaled so as to accommodate most aircraft types without serious limits. It is when operating off the main 'trunk' routes that the problem is more likely to be encountered.

Finally, under the general heading of route monitoring, should come some reference to the checking of airport facilities. Where an operator is already using an airport he will, of course, be well aware of its facilities. But to know one thing, one day, does not mean that changes – for the better or worse – will not take place, and it is in the operator's best interests to ensure that he knows about them in good time. A well run airport will invariably give advance notice of any proposed changes, either through the medium of a Notam or by direct information by means of a letter or telex. But emergencies will occur, and some airports are not necessarily efficiently run, so that no harm is done by keeping a sharp lookout for any changes, or potential changes.

We ought to try to define, or list, what comprises airport facilities. We have already discussed the runways, but there are many other facets that make up the day to day running of an airport, and these *must* be monitored. If a new type of aircraft is to be introduced then the operator must ensure that the airport has the appropriate facilities or equipment to handle it. The first item that comes to mind is the availability of fuel and oil, to the appropriate specification as laid down by the engine manufacturer and contained in the AFM. There must always be a supply of such fuel and oil in sufficient quantity – and to an acceptable quality as well. There is little merit in arriving at a 'new' airport only to find that it normally caters for much smaller aircraft and therefore cannot meet the MSF demand. At best, such a situation can involve an extra technical stop for refuelling again; at worst, it can involve a wait – perhaps of days – until sufficient fuel can be delivered. Then it must be established that the airport has suitable ground electrical power supplies, be these by ground power units (GPU) or ring main. Compressed air may also be required, under this category. The aircraft may need to be properly de-iced, if the airport is in a cold climate and there is precipitation.

While there may well be other operational needs, the foregoing will give an example of the kind of things that must be looked for under 'ramp facilities'.

We have not included radio or radio-navigational aids (navaids) because this information can normally be ascertained from the flight guide, e.g. AERAD or Jeppesen, or from the appropriate AIP.

Next must come handling facilities. Has the airport the appropriate ground equipment for the aircraft? Has it sufficient and suitable stairs or steps? Does it have servicing equipment for the toilets? Has it pure water for topping-up the aircraft's drinking water tanks? Is there an acceptable handling agent available, both to look after the foregoing and also to supply catering to the operator's own standards?

There may be a specialist handling company, or a resident airline, or both. And both should be able to offer the necessary handling services; which one to engage is a commercial decision by the operator's commercial or traffic department. Generally speaking, it is *usually* preferable to use the resident airline for a number of reasons. First, their own airline image is under scrutiny, and they should want to protect this. Secondly, being themselves an operator, they understand an airline's requirements first hand. Thirdly, it may be possible to come to a reciprocal arrangement, in which the resident airline at the airport in question may find it advantageous to contract reciprocally with the visiting operator for similar services at that operator's base. Naturally, such an arrangement can be mutually satisfactory, and is very frequently resorted to. But this is not to denigrate a specialist handling agency, of which there are many worldwide and who provide an excellent service. Really the choice must lie on its merits.

Reference must also be made to the airport's terminal building facilities. Can the terminal cope with, say, a Boeing 747 load of passengers, who may all be in transit, or arriving and departing, or a mixture of both? And can it cope if a second such aircraft arrives at the same time? Does it have adequate refreshment facilities for the passengers, and shops – including the passenger's favourite, the duty free shop? All of these things count, even if some may appear to be trivial. The passenger will use the airline that makes his or her journey the most interesting and least tedious.

So much for the subject of 'route monitoring'. The golden rule, in the opinion of the author, is that nothing is lost by regular route checks by the route planning department, if only that, possibly, an impending problem may be recognised before it manifests itself and timely action can be taken to obviate or minimise any ill effects from this.

11:   Route Licensing

This chapter is included for the purpose of giving a broad outline to the procedures regarded as being essential by some states, before a route licence to operate will be granted to an operator. This licence must not be confused with the operational approval that is necessary before an operator may carry out flights for hire or reward, be this the operation of scheduled services or charter flights. Where *route* licensing is required, this relates to an operator's financial stability and resources, the demonstrated need for the route service, the financial feasibility of this service, and so on. The operational approval of an operator is carried out by a state's regulatory authority to ensure that the operational and maintenance standards of that operator meet the required standards of the regulatory authority. Contrary to many, uninformed, areas of opinion, there should be no difference between the standards of a scheduled or a charter operator. It may be true that, in some examples of the latter, cost-cutting takes place to reduce the charter price, but there is nothing wrong or dangerous in this. A typical example is that in which an operator does not use covered 'airstairs' to the terminal, but parks some way out from this. The passengers will have to walk to the 'gate', but the cost thereby saved shows up in the price of their tickets!

Some states, e.g. USA, have totally 'deregulated' airline services, and operators are permitted to use their own judgement as to whether or not a service will be viable, and at what fare, and what frequency. If it is believed that the service will be profitable then the airline will operate it on simple commercial considerations. But it will still have to meet the appropriate financial stability and safety standards of the state authorities. But, although deregulation allows this freedom of action to suitable operators, it is almost invariably confined to internal services. International airline competition is strictly regulated, and the fare structure must be approved by all of the states to which an operator wishes to operate. As it is normal for such states to automatically have reciprocal rights to parallel services, e.g. where a US operator is granted rights to operate from, say, New York to London and return, a UK operator has the right to operate from London to New York, at the same fare.

Thus the New York-based airline, or, if not based actually in New York, then US-based, must charge a fare that is acceptable to both the UK and the USA governments, and thus becomes regulated. Should a

UK operator choose also to operate the same route then that operator will be required to charge a similar fare or, if no reciprocal rights are taken up, the US operator may, by agreement, operate reciprocal services for the UK operator. Things are not quite as simple as they seem, however. In practice there is no such thing as a standard New York–London, or London–New York fare. Fares vary according to such things as season, time of booking prior to the date of departure, no booking at all with no guarantee of a seat, time or duration of any stopover, standard of cabin service (a few years ago a UK airline, now defunct, operated a no booking service from London to New York where even in-flight meals had to be paid for as an addition to the fare prior to departure, or one could take one's own food aboard. But the fare was absurdly low), and so on.

A final word of caution – deregulation can mean different things in different countries; it is usually seen as a minimum of state commercial interference.

Fares Agreement
Outside deregulated areas or states, it is more normal for the route licence to specify the fare structure, although the operator may well have proposed the fare anyway. But, before the service may be operated, the fare must be agreed by the state of departure and of destination, or destinations. This must be negotiated by the operator, who is required to submit fare proposals to the authorities concerned and who will then signify either their approval or otherwise, in writing. Only when reciprocal acceptance of the fare(s) has been reached and the licensing authority provided with this evidence will the licence be issued.

The Licence Application
Most states have their own rules for the procedures to be followed when an operator applies for a scheduled service licence. These can, and probably will, vary according to whether the service proposed is internal or international. Another factor is the position of the outside state's own national airline, which may well have a monopoly. In this case it may also have an understanding, or something a little stronger, with the national carrier of the state of the applicant. In such cases it is not unknown for both to oppose an application, even if neither wishes to operate the proposed route. A typical argument used in such circumstances is to attempt to show that there is no need for the proposed service. This *can* simply mean that the two national carriers do not wish to be bothered with the route! But, at the same time, they do not wish for an interloper near at hand. After all, the interloper, if his project is a success, may well become a threat to the chosen ones! And merit is not necessarily the main

criterion; politics, not surprisingly, can exert their influence and for no clearly definable reason. A fairly recent example can be found in the UK, as at the time of writing, that is. A large UK airline, as regards scheduled services at least, had applied for, and was granted, a series of licences to a number of new destinations. This meant that the competent authority, namely the UK CAA, was satisfied that not only was there a need for the new services but also that the airline was financially sound and capable of operating these services as detailed in the application. But in the UK the 'competent authority' is also answerable to, currently, the Minister of Transport. And, for no fully acceptable reason the Minister revoked these licences. It may, or may not, have been significant at the time that the UK's largest airline was state-owned, but was in the process of being sold to the public. Whatever the reason for the minister's action, it can only have made the purchase of shares in the denationalised state airline a lot more attractive if the competition had been reduced – or, as proved to be the case soon after – eliminated.

Let us now briefly look at the mechanics of licence application in a somewhat generalised form, *based* on the normal UK procedure. The first thing that needs to be done is for the applicant to initially satisfy itself that there is a real need for the proposed service(s) and that there is a satisfactory element of commercial viability therein. It is usual to carry out a programme of market research to derive a carefully estimated figure of the number of passengers (or, if a freight operation is planned, the amount of freight) that will use the service. Taking as our example, for simplicity's sake, a single route from 'A' to 'B' and return, it is sensible to find out how many people already travel between 'A' and 'B' and by what means. It should involve checks on seasonal variations, and preferably base the pattern on a monthly basis. Is the present method of getting from 'A' to 'B' inhibiting some element of travellers from bothering to even travel, and if so, by how much?

If initial researches show that there is a significant level of traffic between 'A' and 'B' (and statistical sources can normally quantify this level, and by what means of transport), then more research must be done to obtain reliable estimates as to what percentage of this total traffic would use the proposed new air service between 'A' and 'B', and at what fare. It now becomes clear that, whatever the percentage, some amount is going to be diverted from existing transport services, be these surface or air. The remainder will be created by the operation of the new direct service. It will also be clear that the existing transport services carrying the existing traffic will not like this at all, and will use all legal means at their disposal to thwart the plans of the applicant. 'Material diversion' will be the main cry, and these operators will certainly not take into account the needs, or convenience, of the travelling public. When the time comes, as will be seen soon, they will attempt to deride the applicant's estimates, sometimes even failing to take into account the fact that such derision can thereby even apply to their own operation. However, the needs of the travelling public *are* taken into account in their own interests by the authorities.

Market research will now enter a far more precise phase, but this will be started, normally, after a decision has been taken by the applicant to apply for a licence, detailing the frequency and fare structure. It is prudent that this phase is preceded by the application because the 'opposition' will soon learn about the applicant's plans as a result and may submit either a counter-application or a 'variation' to their own existing, but indirect, licence that will permit them to include *some* direct 'A' to 'B' services. It is far better that the applicant files an application before any counter-application, if only to be able to say that their application alone triggered off the opposition's sudden interest in the travelling public. In the process of market research, the applicant will seek not only to make the estimated traffic (upon which the applicant's own internal decision was taken) something more firm, as regards the basis upon which it was made, but also to obtain active support from as many directly interested bodies that want the service. Normally, travel agents, local authorities, etc. will be contacted and given the applicant's proposals. Local politics will be indulged in – dog will eat dog!

The objective, at the end of this phase, will be to have obtained a clear picture of the likely traffic profile, with seasonal variations (if any), and written evidence of the support available for the applicant's proposals (the support need not be only in written form – if a hearing is involved, personal support by witnesses can be invaluable). The opposition will also have been made aware of the applicant's plans, officially. So the next phase is entered into – the preparation of the applicant's submission in support of his application. At the same time the opposition will be marshalling their own forces, ready to counter the case of the applicant, once known.

The submission can either be a simple document or a heavy tome or tomes, according to the characteristics of the case. It will state the need for the proposed service and will provide evidence to support this. This support should be carefully prepared and based on sound principles and researches. It should show, for example, existing traffic levels on the route 'A' to 'B', and why the new application is an improvement upon existing arrangements, and more acceptable to the travelling public. It must not be couched in vague terms, e.g. 'We *think* that 250 000 passengers *could* be carried per annum'. Instead, something more like this is required: 'Our market research programme, carried out in the following manner, shows that the following authoritative sources state that, in total, some 250 000 passengers will be carried annually between 'A' and 'B', according to the following pattern'. Then should follow a detailed breakdown of the MR programme referred to. To use the vague approach is to court a refusal of the application.

The submission will then show, using the proposed fare structure, and breaking this down into proportions if appropriate, (first, 'executive' and economy classes) the revenue calculated to be forthcoming *from all sources*, should the application be successful. Yes, there are other sources of revenue than from fares alone. If the route is international, and the rules permit, sales of duty free and other goods can provide not-insignificant income. If the applicant has established the necessary agreements with other carriers, he can sell through tickets across those other airlines' route networks, and thereby receive a percentage of those carriers' fare proportion of the whole. This is known as 'interlining'. How much this will affect the applicant's income will, of necessity, depend upon the relationship between the proposed 'A' to 'B' service and the other airline's routes included in the ticket sold by the applicant. If, for example 'A' to 'B' is a comparatively short haul, say, 150 miles, and the through ticket sold by the applicant, including the other carrier's routes, covers a further distance of several thousands of miles, then the commission earned on this element of the total through distance could well exceed the fare from 'A' to 'B'. Again, the proportion of income resulting from these other sources must be soundly calculated from reasoned estimates. But, include *all* possible but acceptable sources of revenue, subject to being able to justify these.

Having shown estimated total net income – preferably on a yearly basis and over a three year period – backed up by a monthly cash flow projection, a not-dissimilar exercise must be carried out showing the costs. Normally these are broken down into three main headings, covering administration and overheads, maintenance, and route costs. The first of these headings comprises all those costs that must be incurred before any flying whatsoever takes place. Under this heading one would expect to find overheads, e.g. accommodation, telephones, postage, advertising, and other miscellaneous charges. Then comes salaries, and the standing charges for aircraft and spares holding. The latter item will vary according to how the aircraft is being acquired. If one assumes a straight purchase then it is normal practice to amortise, or write down, the total cost of aircraft and spares over a stated period of time, normally eight to ten years, and with a notional remaining value at the end, say 20%. For example, assume that the cost of an aircraft with spares, is, say,

$25 000 000. Deduct 20% from this and we obtain the figures $20 000 000. This is the sum that must be amortised over eight years, and the residual value is, of course, $5 000 000. $20 000 000 ÷ 8 = $2 500 000, and this is the standing annual cost of the aircraft. (Naturally, if a fleet is being considered the cost is similar but multiplied by the number of aircraft in the fleet. But the scale of spares held can be reduced on a per aircraft basis. That is to say, the value of spares held for one aircraft will not be double for two, and so on.)

It would be somewhat naive to assume that this is the end of the financial costs involved, even though the aircraft has been assumed to have been purchased outright. To be taken into consideration is what the money tied up in the price of the aircraft, i.e. in this case $25 000 000, could be earning through investment. So a loss on investment allowance should be added to the aircraft costs annually. Maybe the aircraft is being leased – if so, the annual total lease payments must be given, or, if hire purchase is involved, the annual repayments total. Insurance, too, must be included under this first heading. There will be hull insurance – this covers the loss of the aircraft – and a premium to be added, based on the number of passenger seats. Third party cover must also be provided to a sum that is satisfactory to the regulatory authority. Sometimes this last cover causes some head scratching, and wondering what to allow for. The risk to be assumed must be sensibly based and also realistic. A cynical view of the 'worst possible risk' case is that involving a mid-air collision between two fully-loaded B747s, each containing solely the dependants of the US legal profession and taking place over the City of London! There should also be insurance of property and other assets, vehicles, and so forth. Finally, it is worth including an annual contingency allowance. Adding up all of these will give an idea as to the annual costs of overheads and administration.

The next costs heading is that of maintenance and this can also take various forms. For example, the airline can contract-out all maintenance, and thus have a known figure. But this must, naturally, include an element of profit for the contractor, and it is therefore not necessarily the most economical way. The whole question hinges upon the scale of the airline's operation; contract maintenance is best suited to a small operation, not large enough to justify the setting up of a maintenance organisation 'in-house'. The staff salaries, and engineering facilities could well outweigh any contractor's profit. But, for the larger organisation, it is worth undertaking one's own maintenance. This can be carried out on what is known as a progressive maintenance schedule, in which some specified work is carried out daily within a given period, so that the whole, calendar time, required programme is carried out within a year or approved period. Alternatively a cycle of Checks can be assumed, each including a specified work content, and involving the aircraft being out of service during the time taken to perform each Check – unlike the progressive maintenance plan. Each Check must take place on the basis of hours flown or a calendar period, whichever comes first, and some Checks will take longer to perform than the others. Normally there will be a pattern of Checks that runs something like this: Check A–3 months/ 200 hours; Check 1–6 months/1000 hours; Check 2–12 months/3000 hours; Check 3–24 months/7000 hours. (This cycle is given solely as an illustration and may well bear no relationship to any actual programme.)

Engines, too, can be overhauled to differing arrangements, subject to the requirements being met. Every engine has an approved time between overhaul (TBO), expressed in hours. The cost of an overhaul can be ascertained from specialist contractors and the costs per hour become a matter of simple arithmetic, namely, the overhaul cost ÷ TBO. But it is better to assume that no engine is going to actually *achieve* its approved life, and therefore to assume a percentage reduction in the TBO figure for financial purposes. Or, if preferred, the airline can enter into a contract for engines based on 'power by the hour', i.e. it only pays a fixed rate, per hour flown, for all engine maintenance.

In a case such as currently being discussed, the maintenance staff salaries and overheads will be included in the first heading. The engineering costs will therefore comprise such things as the cost of rotables – items that have to be replaced on a time basis – materials, and engine overhaul costs. There will also be radio and avionics, and instruments overhaul costs, and these will again depend upon whether the airline carries out its own avionics maintenance or operates on an exchange basis. But the foregoing will give an idea as to what may be involved.

The final heading is route costs. These are the actual costs incurred in each flight, between 'A' and 'B'. First there is the cost of fuel and oil consumed, from starting engines to shutting-down. Then there is a landing fee charged by airport 'B' on arrival, and 'A' on return. If the operator does not have staff at 'B' then a handling fee will be payable, either to another airline or to a specialist agent. There may well be a route navigation charge, and possibly passenger taxes, *pro capita* – although the latter item can be passed on to the passenger direct, through the ticket. And the crew will have to be paid an hourly flight allowance.

Incidentally, some allowance should be made for so-called 'dead flying' and this covers all non-revenue flying such as training, positioning, air tests, and the like. The cost cannot, of course, be established until the total hourly cost of the aircraft has been established, as outlined above.

If this is the applicant's first route, the route pattern proposed is known, and therefore an annual utilisation can be established. The total annual costs are known and so the cost per flight hour can be calculated; this includes the entire operation, namely operating the route 'A'–'B' × times per annum. If, however, the applicant is already operating another route or routes then only a proportion of the total non-route costs need to be taken into account, on the basis of the total flight time for 'A'–'B' against the airline's annual total for all routes.

In the submission we now have a schedule of estimated revenue and schedules of costs expected. In an ideal world the former should be greater, but if this is presented as being the case for the first year, at least, the application may not be believed and will probably be rejected. An air service will not develop its full potential in its first year of operation. It should not be forgotten that the great bulk of the estimated traffic has become used to going the alternative way and must be attracted away from it. This takes time, and the first year of operation will almost inevitably be operated at a loss (or, it must be assumed to). The expected, and average normal pattern is for year 1 to run at a loss, for year 2 to just break even, and for year 3 to result in a small profit. Therefore, it is as well to include a three-year projected cash flow in support of the application to show such a trend. Year 4 and on is where the profits are made! (Hopefully.)

It should be stated here that the foregoing represents the costs or pattern of no particular operation or aircraft, neither does it represent the requirements of any particular country. It is simply a guide as to a typical data format that comprises a submission in support of an application. It is normal practice for the applicant to be required to provide all parties that have declared any form of interest in the application, whether they be objectors or interested parties, with a copy of the submission.

The Hearing
We have now reached the stage at which both the application and the submission have been filed, according to the requirements of the state in question. All the so-called 'interested' parties have been notified, and, where appropriate, have been supplied with a copy of the submission, which will contain not only the applicant's forecasts of revenue and costs but also evidence of support. The next stage that will take place (in the UK, anyway) is that a private meeting between the authorities and the applicant will be held to examine the applicant's financial position, but, if the applicant already holds a licence or licences, this can be dispensed with. The object of such a meeting is to protect the travelling public's interests, in that the authorities try to establish beyond reasonable doubt

that the applicant is well-heeled enough to survive the process of starting an air service, and of continuing to operate it, without becoming insolvent and thereby either strand passengers or have them forfeit their ticket money. In the past, the sudden demise of a number of large airlines has resulted in thousands of would-be passengers being owed many thousands of dollars, or pounds, in respect of pre-paid fares, thus becoming creditors of the fallen airline. The private meeting's proceedings are confidential.

A date is then set for the hearing, and this bears some similarity to a court of law although far less formal. The applicant will present his case, and produce evidence and/or witnesses in support. Both the applicant and these witnesses can then be cross-examined either by the authorities or the opposition. Then the opposition can submit their case, giving their reasons and, if necessary, producing their own witnesses and evidence. Here the applicant and the authorities can cross-examine the opposition, in turn. A summing-up process then follows, after which it is normal for the authorities to close the hearing and go away to think and deliberate on what they have read and heard. After a few weeks their decision is then delivered – yes, no, or 'maybe, but if'. If the result is 'yes', the applicant can go away and start the new service. If 'maybe, but if', then the applicant has to meet certain specified requirements, after which the decision becomes converted to 'yes'. Usually the 'maybe, but if' decision is based on a requirement for the applicant to make certain alterations in his financial structure or something similar. If the decision is 'no' then the applicant may well have the right to appeal against the decision; if 'yes', so may the opposition. But, assuming that the final outcome of all the processes results in a 'yes' then the applicant can go out and, armed with a licence, start to operate the new service in accordance with the terms of this licence.

Note
It must be emphasised that the requirements and proceedings for the obtaining of a licence vary from state to state. While the foregoing is *based* on the UK requirements it is in no way a statement of these and should be taken as simply an illustration of what may be the case, depending upon the state involved. The whole process may be easier, or more stringent, according to the state's view of airline operations, competition, and the like.

12:   Aircraft Evaluation

The reader may be familiar with the expression 'horses for courses', and this applies equally within the air transport industry, except that the means of transportation is metal and carries more than one person. But it is important that the vehicle meets the requirements and demands of the course, or courses, that are involved. The process of aircraft evaluation must, therefore, be carried out in a totally objective and impartial fashion, and the sole criterion is – does the aircraft meet the requirements of the route, or route pattern, or could another type perhaps be the answer? And, of course, is there any merit, in terms of profitability, in changing from the existing aircraft, if any? The following pages will attempt to describe how an aircraft could be impartially evaluated by route planning.

Assuming that the airline is already a carrier, the most probable reason for considering re-equipment is growth and/or efficiency, or competition. If the carrier is a fledgling and an aspirant it may be wondering what is the optimum aircraft type to start with is. But in both cases the real question is the same, namely, what is the best aircraft for the job? In the beginning it will almost certainly be management or commercial who start the process by presenting to route planning the requirement and asking for recommendations. Even at this point there can well be a lack of clarity, for example, the whole question can be variable and therefore difficult to illustrate. It is felt that perhaps the best way to choose a typical example is to quote from the author's own experiences, and in particular one illustration.

The Competitive Evaluation

Let us assume that the existing route network is known and defined. Some stages may have been added, and the existing aircraft are becoming obsolete and unsuited to the new stages. With what aircraft should the operator re-equip? The first process is for the entire network to be considered in toto. Stage or sector distances must be measured, and the airport runway lengths checked or ascertained. This process must also include the consideration of alternate airports. Also to be verified are the statistical meteorological conditions, such as en route winds, temperatures, special weather 'freaks' or manifestations, and such matters. If the necessary airport data is not available in the proprietary flight guide then the appropriate national Aeronautical Information Publications (AIPs) will usually provide this, and can be consulted (if not already held by the operator) through the nearest national Aeronautical Information Service (AIS), who will normally either provide, or sell, photocopies of the relevant pages. Airport facilities and fuel availability must also be ascertained; if there is no fuel of the right specification available, suppliers in the operator's own country must be contacted and asked if they would be prepared to make such fuel available. If so, the annual estimated uplift must be communicated to them once known in order to obtain a price.

We now know both the form of the task, and have the necessary information to study this in more detail. The next thing is, what required aircraft characteristics are to be dictated by the airport and meteorological conditions, and also by the commercial requirements. For example, there is little merit in looking at, say, a B747 when the commercial requirement is for a maximum of 100 passengers (pax.) per trip. Any route planning organisation worth its salt will know what 100 seater aircraft are available on the market, and their outline operational characteristics, such as runway requirements, the effects of temperature and altitude, and so forth. If this has not already been done, send to the aircraft manufacturers for their 'brochures', which will be of use in narrowing down the field of contenders without a lot of hard work. Let us assume that from a study of the available brochures we have been able to make a shortlist of three aircraft, each of which would *appear* to be suited to the task. We will refer to these aircraft as 'X', 'Y', and 'Z'; from these must be selected the best for the whole task. An inevitable complication that can set in now is that each may be better on certain sectors than the others and so it must be the overall result that counts. This must be a commercial decision; route planning must simply provide data, and operational recommendations, after carrying out a comprehensive analysis of each aircraft across the whole network. To do this the brochures must be put on one side and more formal documents obtained from the manufacturers. Flight manuals and cruise control manuals are required, plus weight statements.

It is probably best if we first of all take a look at the weight statements. These should be based on aircraft actually flying, if possible, and not sales material. The latter is prone to optimism! So, we have three different types of aircraft, each of which appears to be at least suitable for the route network to be considered. But, to do a fair analysis, there should be the greatest possible degree of commonality between these aeroplanes. First of all, in conjunction with the weight statements, find out exactly how much equipment is installed to customer's option. (The weight statement is a document that shows in some detail how the empty weight of the aircraft is made up.) Ideally all radio and avionics, furnishings, seats, and galley equipment should fall into this category. Now, on paper, strip the maximum possible from each aircraft, taking care that each item removed from one is also removed from the others. In a few instances this may not always be possible and such items must be regarded as being part of the basic weight.

The Basic Weight

So far as is possible we now have each aircraft without radio, avionics, seats, galley equipment, and the like. The next step is to replace everything that has been removed, having first of all deducted the weights of the items removed. But in replacing these items we are referring to their function. That is to say, we re-equip each aircraft with the operator's own standard 'fit'. This can be either a preferred choice or a scale common to other aircraft that he may be operating, in the interests of economics and experience (e.g. spares holding considerations). Re-equip each aircraft as far as possible, identically; this operation is best carried out in consultation with the engineering department, and preferably with the weight and balance engineer participating. At the end of this exercise we will have the aircraft under consideration all equipped with the same radio, avionics, autopilots, flight instruments, flight management systems, seating, and galley equipment. The weights of all of these items must be added to the basic empty weight for each aircraft type, so that we have once again an operating weight empty (OWE) for each, but this figure will be such that each type has been more fairly treated for our purposes by bringing in the highest possible degree of commonality. And, as regards seating, try to ensure that each aircraft has the same number of seats, if at all possible. This is because at some stage of the competitive evaluation the financial people will need to know the seat/mile costs. If, in fact, the type that is finally chosen can carry more seats, then this can be a bonus. But our present task is to try to establish which type is best suited for the route network, and we have assumed that it will have 100 seats, as indicated by the commercial department. If one type can only offer, say, 98 seats, however, then *assume* that the other two are also 98 seaters.

Having now established OWEs for each aircraft it becomes necessary to convert these numbers into aircraft prepared for service (APS) weights. The APS weight is the OWE, including unusable fuel and oil, to which has been added the weight of the crew and flight equipment (such as manuals, and navigation equipment not forming a part of the equipped aircraft), water (galley and toilet(s)), passenger service items, bar boxes, duty free goods (if applicable) and catering (in-flight meals, to an appropriate scale). The difference between the APS weight and the OWE for each type should be the same, if the foregoing processes have been 100% successful in the achievement of commonality. A difference of a few kilogrammes need not be a source of concern.

Now a position has been reached when the maximum payload for each type can be calculated – if a single act of simple arithmetic can be dignified by the use of the word 'calculated'! The maximum *possible* payload for each aircraft type is simply the certificated maximum zero fuel weight (MZFW) minus the APS weight. The MZFW is defined as the maximum weight permissible, above which any weight increase may only consist of fuel. It is worth calculating, at this stage, the maximum amount of fuel that can be carried with the aircraft at MZFW. Later we can use this to establish the payload/range capability for each type. The maximum fuel weight for MZFW is simply the maximum takeoff weight authorised (MTWA) minus the MZFW. Having reached this point we are getting some indications already as to comparative capability, although these are not yet very significant at this stage, except for the maximum payloads, which we shall use later.

Incidentally, a brief note here about maximum payloads, limited by MZFW, may not be amiss. We are discussing everything, up to the present, in terms of *weight*. But there are some transport-category aircraft that cannot carry their MZFW-limited payload because of internal dimensions, particularly when considering passenger aircraft. In such cases, the payload becomes volumetrically limited, or to use a colloquialism, the aircraft 'cubes-out'. This is a function of the density of the payload – a cargo of gold, for example, would be limited by MZFW, while a cargo of, say, fruit or vegetables would almost certainly be volume-limited (in the case under discussion). However, we are evaluating a nominal 100 passenger capacity trio of aircraft types and will not digress further.

Now turn to the performance section of the AFM for each aircraft type, and calculate the maximum takeoff weight (MTOW) and maximum landing weight (MLW) for every airfield and runway in the network, including the alternates. These calculations must be made as appropriate to the statistical temperatures likely to be applicable for the operation, working seasonally, if appropriate. Due account must be taken here of the effects of airport altitude amsl and thus of WAT. (Landing WAT is only accountable as regards the effects of altitude in ISA.) Where WAT becomes a limiting factor, does any of our trio have certificated procedures for minimising the effect by increasing V_r ('overspeed' techniques)? If so, this must be taken into account when calculating maximum TOW for any airports that are WAT limiting rather than runway.

We now have a list of maximum TOWs and landing weights for the network. But we now need *required* weights, rather than RTOWs, which can be somewhat academic. The following procedure is one way of dealing with this, and it is well worth bringing the commercial people into the scene for this purpose. The main question to be asked is: do the commercial department want, essentially, the maximum possible payload (by weight), the maximum possible numbers of passengers, and if so, what passenger unit weight do they wish to be used? Internationally, there are standard weights for passengers, viz: males 75 kg, females 65 kg (both being over 12 years of age), children 40 kg (under 12 years but over 3 years of age), and infants, who are deemed to weigh between 8 and 10 kg, according to the route. Under UK regulations, children apparently weigh less if *not* travelling between the UK, Channel Islands, and the Isle of Man. Under the Air Navigation (General) Regulations 1981, a child within the region mentioned above weighs 40 kg, and an infant 10 kg. But step outside these boundaries and they shed a kilo – a child weighs 39 kg and an infant 8 kg.

The weight of the passenger's baggage is required, also, as this forms part of the eventual unit weight. Under the above UK regulations this varies with the route, thus: UK internal, cabin baggage – 3 kg, hold baggage – 10 kg (13 kg if on a holiday route). European journeys increase a little as regards hold baggage weights, these being 12 kg (13 kg on a holiday route), while for intercontinental routes, the hold baggage weight is 14 kg (but 16 kg if on a holiday route). These are, like the passenger weights, standard. On some routes, notably those connected with the North Atlantic, there are volumetric baggage allowances, but it is not our worry at this stage to further complicate matters. Our immediate problem is to establish an average passenger unit weight – that is, the weight of a typical passenger and his or her baggage. A weight of 85 kg has been used in the past, and it is proposed that this is now assumed for the purposes of our study. As we are working on an assumed 100 pax. capacity this means that the *required* sector payload if 8500 kg, for our purposes anyway.

The Required Takeoff Weight

We have now established a notional required route payload. Adding this to the APS weight will give the required zero fuel weight (ZFW). To conduct the evaluation we will employ the 'payload loss' method – i.e. the reduction, if any, that needs to be made to the required payload (viz: 8500 kg) in order that any overriding limitations or operational requirements be satisfied. The first of these is to ascertain the required zero fuel weight, which is simply the APS weight plus 8500 kg payload. Let us assume that the APS weight is 21 950 kg, and that the payload is, as discovered above, 8500 kg. The required ZFW is, therefore, the sum of the APS weight and the payload, which equals 30450 kg. It must be remembered that limitations and operational requirements take precedence over payload and that if any values arise that exceed these weights then *only* the payload may be reduced in order to achieve compliance.

We have already calculated the RTOW and LW requirements, as imposed by either runways, or altitude, or both. We know the required ZFW and payload. However, one important item is missing, namely the sector fuel requirement. After calculating this, as described in Chapter 7, we can add the MSF so calculated to the required ZFW and thus obtain the required TOW. So, for each aircraft type, and for each sector, we do this and arrive at overall route network required takeoff weights, sector by sector. We know, too, the maximum TOWs and LWs. And, in the process of calculating MSF we also know the expected sector burnoffs. Compare, for each type of aircraft (X', 'Y', and 'Z') the required TOW against the RTOW, for each sector. If the RTOW exceeds the required TOW then the takeoff does not restrict the payload. But, if the reverse obtains, then the payload available must be reduced so that the TOW does not exceed the RTOW. For example, assume that for the sector 'A' to 'B' the MSF is 6,050 kg. The required ZFW is the APS weight 21950 kg plus the required payload 8500 kg, which equals 30450 kg. To this required ZFW must be added the MSF 6050 kg, and this gives the *required* TOW 36500 kg. However, because of airport elevation and temperature, i.e. WAT, the regulated TOW is 36250 kg. We thus have to reduce the weight by 250 kg for the aircraft in question, and this can only be done by reducing the *available* payload by the same amount, so that the sector payload is 8250 kg. Assuming that this result was obtained using aircraft 'X', repeat the process for aircraft 'Y' and 'Z', thus finding the one that offers the highest payload. And, as an important part of the process, when obtaining the MSF also obtain the block times and burnoffs.

Taking this one sector we have now established the regulated TOWs for all three aircraft. But we must check that there are no problems as regards the landing weight at 'B'; this is unlikely, but check nevertheless. The regulated landing weight may not be less than the RTOW minus the burnoff; if it is, then the payload must be still further reduced so that the actual TOW does not exceed the RLW plus the burnoff. Again, the process must be repeated for all three types of aircraft being considered. It is suggested that three columns be drawn, one for each type, and the sector payload, block time, and burnoff be tabulated vertically, with MSF, payload loss, and limiting factor, against each sector horizontally. At the end of the exercise we will have a clear indication of each aircraft's cpabilities, on equal terms, as applied to the route network. The choice must then be made on considered commercial terms, after further practical work by route planning.

Table 12.1 Aircraft evaluation – order of differences.

Leading particulars	Aircraft 'X'	Aircraft 'Y'	Aircraft 'Z'
MTWA (kg)	40 000	40 500	39 500
MLWA (kg)	35 500	35 000	35 000
MZFW (kg) (a)	32 800	33 000	31 850
APS Wt (kg) (b)	21 950	22 875	21 522
MZFW payload: a – b (kg)	10 850	10 125	10 328
Configuration	100 Y	100 Y	100 Y
OWE (kg)	19 210	20 135	18 782
Max. fuel: – brake-release (kg)	10 000	9 150	8 900
Standard reserve fuel (kg)	1 550	1 620	1 525
Max. fuel, less reserves (kg)	8 450	7 530	7 375
Still air dist. with max. fuel			
(nm/block time (hr.min)	950/2.38	910/2.35	870/2.30
Payload with max. fuel and			
at MTWA (kg)	8 050	8 475	9 078
Fuel at MZFW (kg)	7 200	7 500	7 650
Still air dist. with max. payload			
(nm/block time (hr.min.)	715/1.58	710/2.02	510/1.28

Notes:
(1) Aircraft 'X' has highest MZFW payload, best distance with full tanks, and with max. payload, but has lowest payload with max. fuel.
(2) Aircraft 'Y' has lowest ZFW payload, second-best distance with full tanks, and with max. payload, but higher payload than 'X' with max. fuel.
(3) Aircraft 'Z' has second best MZFW payload, lowest distance with full tanks, but highest payload with full tanks.
(4) All conditions assume no restrictions on MTWA. Distances are in nm, weights are in kg. Configuration Y = economy seating.
(5) The standard weight of customer options, to company standard, is 2740 kg.

Table 12.1 shows how three aircraft, all within the same category, compare when initially examined. In each case each aircraft has been stripped of all the manufacturer's standard equipment (as defined earlier) and has then been re-equipped to the operator's own standard. The APS weights quoted reflect this element of commonality. The results show that aircraft 'X' can carry the highest payload permitted by structural considerations (10 850 kg), and can carry this for a stage distance of 715 nm, assuming a standard diversion distance of 100 nm with a fuel contingency allowance of 5%, and 30 minutes holding at 1500 ft above the alternate airport. This is also the best aircraft for carrying the maximum payload the greatest distance. But with the maximum fuel available at brake-release (i.e. full tanks on the ramp) it has the lowest payload (8050 kg). Aircraft 'Y' has the lowest structurally-limited payload (10 125 kg), the second best payload with maximum fuel (8475 kg), and the second best distance with full tanks, (910 nm). Its still air stage distance with maximum payload is also second best, at 710 nm. Aircraft 'Z' has the second best structurally-limited payload (10 328 kg), the highest payload with maximum fuel (9078 kg), but the lowest stage distance with maximum brake-release fuel (870 nm). Its still air stage distance with maximum payload is significantly the lowest of the three (510 nm). It can carry more fuel when at MZFW, but burns this to achieve the lesser distance. On most counts, with the exception of the MZFW limited

Table 12.2 Comparative route analysis – stated payload requirement.

From	Sector To	Dist nm	Block time	MSF kg	B/off kg	P/L loss	P/L kg
				'X'			
'A'	'B'	500	1.44	6050	4450	250	8250*
Alt.	'C'	105					
'B'	'A'	503	1.39	5800	4200	nil	8500
Alt.	'D'	98					
				'Y'			
'A'	'B'	500	1.45	5790	4150	375	8125*
Alt.	'C'	105					
'B'	'A'	503	1.40	5670	4381	nil	8500
Alt.	'D'	98					
				'Z'			
'A'	'B'	500	1.40	5960	4250	127	8373*
Alt.	'C'	105					
'B'	'A'	503	1.36	5750	4050	nil	8500
Alt.	'D'	98					

Notes:
(1) This analysis is based on stated required payload of 8500 kg. Except where a 'P/L loss' is noted the *available* P/L may be higher than 8500 kg.
(2) Burnoffs and MSF have been calculated using 85% statistical winds. Block times are given for 50% winds, and include 15 minutes taxying time (10 min out, 5 min in).
(3) All weights are in kg, and times in hours and mins.

* Denotes WAT limits. Ø Denotes takeoff runway limits. L Denotes landing runway limits.
   (Only one return sector has been analysed as an illustration, and in this case only WAT limits on the outbound stage.)

This illustration is based on no known aircraft types but is considered to be representative of a certain category of aircraft. Ø and L are not involved in the above example.

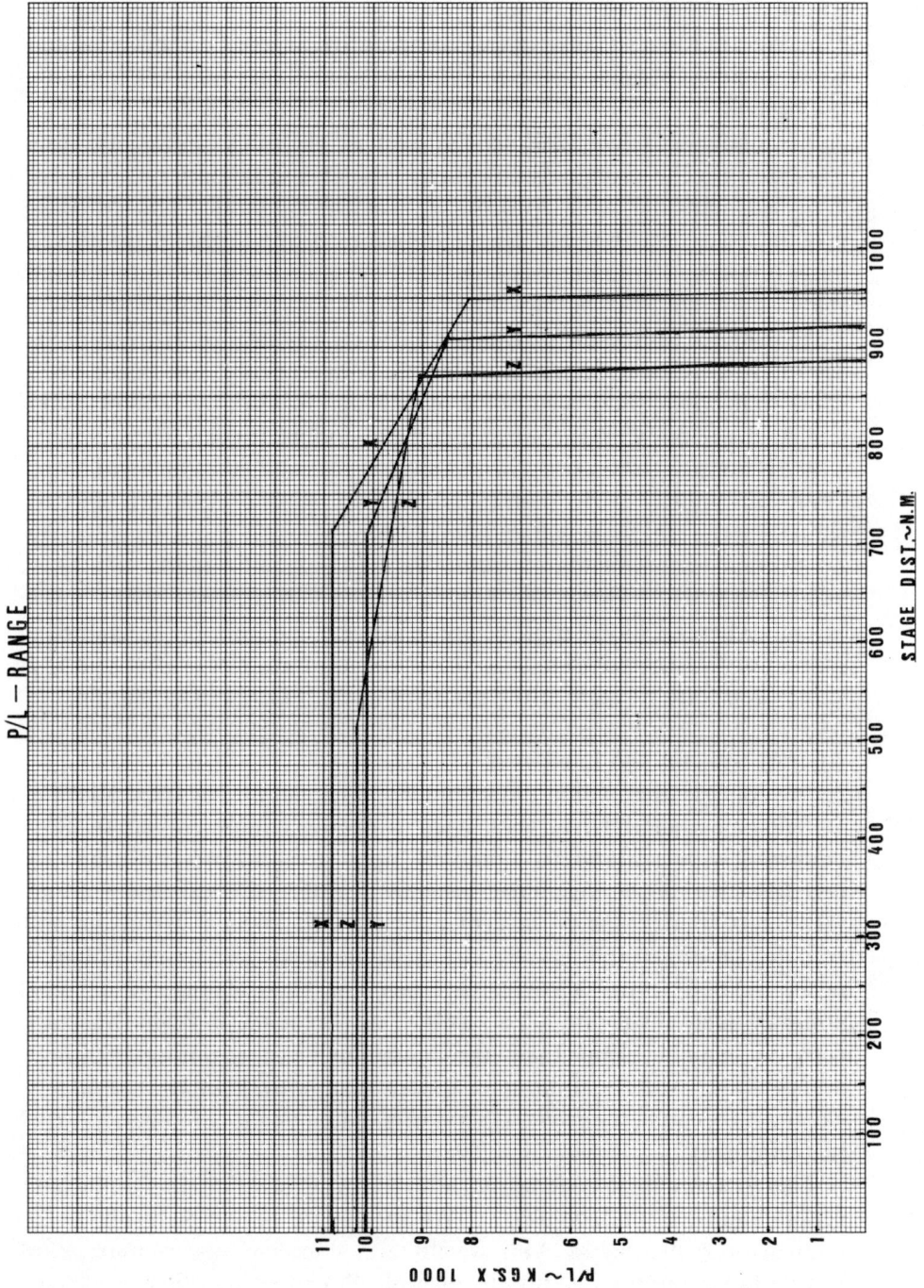

Fig. 12.1 A payload-range comparison of three different aircraft, as described in the text. This clearly shows how the various points, for and against, emerge.

payload, this aircraft does not appear to have much to commend it. However, in the next stage of our comparative analysis we will continue to examine it, because some commercial advantages may yet manifest themselves.

If we refer to Table 12.2, we can see how the three aircraft are now looked at on the route network; for simplicity's sake only one route is analysed, but the same process is used throughout. The MSF is established as described previously, in each case, and the block time is the stage time plus 15 minutes taxiing. The payload is the RTOW, minus the APS weight plus MSF *or* the MZFW payload, whichever is the lower. The stage distance is, in each case, taken as being a typical short-to-medium sector, with normal diversion distances. The workings are not shown, other than where these have some overall significance; the final decision will be made on what the aircraft can do, and what costs are involved. There is nothing to be gained in introducing such things as the flight levels assumed, runway performance, and so on, as such matters tend to obfuscate the real issue, which is solely commercial – but subject to any operational recommendations that may be valid and made at a later stage. Figure 12.1 shows the payload-range performance of the three aircraft schematically; it must be borne in mind that none of the examples given purport to represent any particular aircraft types, although the values given are typical for the group of aircraft under examination.

Aircraft Systems and Equipment

In the comparative evaluation of the three aircraft as discussed above it will be remembered that all three were stripped of equipment and then re-equipped with a 'standard' company pack. It is not inappropriate to comment on this at this stage, because if a new aircraft type is to be introduced it might just as well enter service with the most useful and up-to-date equipment that is suited to the route needs and is acceptable in costs. Indeed, some systems can pay for their capital cost in a comparatively short time, as a result of savings made in flight time and fuel consequent upon their use. (Although, as mentioned earlier, compatibility with the equipment already being carried on the operator's existing fleet can be a powerful factor in deciding on the equipment for the new aircraft.)

The first main consideration, after bearing in mind the minimum requirements, is that of weight. Weight saving here will result in an equivalent reduction in APS weight and consequently in maximum payload or economy fuel uplifts, or both. But there are other factors as well that should be considered and these have very important commercial considerations. For instance, does the proposed re-equipment on the route network appear to benefit from the introduction of Category 2 and 3 AOM? In other words, is schedule-keeping of prime importance to the travellers who are expected to use the services? Or are the services essentially 'holiday' routes, and thus more able to tolerate delays due to weather? If Cat. 2 or 3 operations are to be considered then it should be borne in mind that the aircraft *must* be fitted with Cat. 2 or 3 ILS (instrument landing system) equipment, and this includes autopilot installations and even auto-throttle. By the same token, the ground installations must also be compatible with Cat. 2 or 3 requirements. There must be a high level of accuracy, and fail-safe characteristics, both on board and on the ground. And the ground ILS *must be* of a 'no break' nature, i.e. should the installation fail, a standby must cut in automatically with effectively no measurable interruption of guidance.

Having decided this first equipment question, and assuming that it is in favour of Cat. 2 or 3 operations, then it is sensible to assess if further additions to this installation are necessary. It is almost certainly less costly to initially instal a complete package rather than to retro-fit additional aids. So, having decided on Cat. 2 or 3 operational capabilities for the new fleet, shall these aircraft be equipped also with computerised flight management systems (FMS), which optimise the flight procedures for each stage – e.g. selection of optimum FL and engine power settings – engine pressure ratios (EPRs). Although this may appear to be a never-ending list, it sounds worse than it really is! But why spoil the ship? With such a scale of avionics it might be thought that enough is enough. But, why not go for full automation at this stage, instead of only, say, 80%? Add to the FMS an interfacing area navigation system, such as Omega.

The Automatic Sector
The use of area navigation systems has two main advantages, and we will take as the first, the integration with the already-discussed systems. The normal *required* avionics carriage for airways flying is twin VOR and ADF, i.e. two of each. VOR stands for VHF omnidirectional radio range, and ADF for automatic direction finding equipment. Both of these systems are based on point source ground equipment, in that both VOR and ADF provide a bearing from or to, a given ground facility (VOR can also include distance measuring equipment (DME), in which case it provides both a bearing and a distance from the facility). If the aircraft is only fitted with these point source aids it will normally have to fly from point to point, within the range and capability of the aids being used, and this implies that airways or advisory routes (ADRs) will normally need to be followed. But, if an area navigation system is carried, quite long 'straight line' en route clearances can be forthcoming – often referred to

as 'direct to' clearances. Normally these clearances are from one point source aid to another, but the distance between them can be several hundreds of miles, and this may result in a significant saving in distance. (As mentioned earlier, the author has seen a 'direct to' clearance issued to an Omega-equipped aircraft from overhead a VOR near to Nuremberg to a similar facility near Ostend – a distance of several hundred nautical miles.) With a fully-integrated and coupled installation Omega can, through the FMS, not only navigate the aircraft along the new, direct track, but can also optimise the engine thrust, through the auto-throttle, to suit the revised ETA, and thereby save fuel. So a fully-integrated avionics fit can provide reliable schedule-keeping (in comparison to a less-well equipped aircraft), *and* it can reduce the burnoff significantly.

Other equipment scales do not really involve route planning, as these will be essentially confined to such items as the type of seats to be fitted, galley equipment, passenger service items, catering and refreshments, and the like. But, once route planning has issued its report and recommendations, it is well worth while holding a conference of all interested departments before reaching a final decision.

Once the desk studies of the three aircraft have been completed, but prior to the final selection of the aircraft, it is most advisable that all three should be flown by the operator's training captain designated for the new fleet, and the flight technical personnel from flight development. The main objective here is to check on handling characteristics; a prime consideration will be assessing the conversion training required, although the initial conversion and type rating should be carried out at the manufacturer's technical flight school. But the operator may have some 'difficult' airfields, and handling characteristics can then become important. What type of route training and checks are going to be involved? Are these already regulated by frequency and time intervals? How much training flying is likely to be required? How does the aircraft handle in gusty winds or crosswinds? And so on and so forth.

Thus we reach the end of the route planning contribution to the evaluation; as can well be imagined, it is a considerable one. On the face of it, aircraft 'X' appears to be a clear winner. But aircraft 'Z' should not be totally discounted, because on the shorter sectors it can carry a greater payload than 'Y', and at one point, namely 870 nm, it carries the greatest payload of all three. But it has the shortest stage capability of them all. Figure 12.1 shows the three aircraft's payload/range capabilities diagrammatically, when carrying the same reserves. But note this – these reserves do *not* comprise the same weight of fuel but are for the same diversion distance, contingency, and holding time. While route planning has been evaluating the aircraft operationally the engineers will have been comparing maintenance implications and costs, while commercial will have been undertaking a study of the market and assessing what is needed, in terms of aircraft size and characteristics, to meet this market. To be dazzled by high payload/range figures may not result in the optimum choice; if the passengers are not there a number of empty seats will be, and these cost money to carry. Again, there is usually a reasonable amount of charter work available, and the potential must be assessed and applied to the shortlisted aircraft. If there is a considerable indication of charter potential this could help to outweigh other considerations, for the choice will be made, in the end, on market forces and operating costs, and just possibly on customer appeal, plus, lest it be forgotten, even if the passengers are not there in sufficient numbers to totally fill the aircraft, any spare, or residual, weight availability can be used to carry extra fuel. If this capability is in evidence at airports where fuel is cheap then it can be turned to profitable effect – i.e. load up with cheap fuel.

13:   Route Proving

In many walks of life it is quite normal for some kind of checks to be made to ensure that things fit together as they were intended to. A cabinet maker will see that his joints result in a drawer fitting into a cabinet; a motor-car maker will go out of business rapidly if the doors of his cars are smaller than the 'holes' that they are intended to close. The doctor of medicine will be called before his (or her) peers, if the treatment prescribed directly causes the patient increased ill health, or worse. Although, admittedly, the very occasional exception to the rule does occur, the precaution common to all such situations mentioned above, by way of illustration is: check for fit and/or suitability. This is the function of route proving when applied to air transport by airlines.

What is Route Proving?
In the preceding chapters we have gone through the various functions of what constitutes route planning, flight documentation, route licensing, and aircraft evaluation. We have now reached the point at which the new route licence(s) have been issued and the new aircraft have been acquired. Now the crews for these aircraft need to be trained and converted on to the type before any fare-paying pax. can legally be carried. The training captains will have arranged with the aircraft manufacturer beforehand for nominated pilots, and, if applicable, flight engineers to have attended a technical course that will enable them to take the technical examination relating to the aircraft. The manufacturer will also, probably, give complete training, including flying, that will enable the aircrews taking the course to leave with a type rating appropriate to the aircraft included in their licences. On the other hand, the airline may carry out its own training programme, subject to having the appropriately qualified staff to undertake this. For example, the flying training and type ratings must be carried out by approved type rating examiners (TREs). So somebody, at least, has to go on a suitable course, somewhere. If the manufacturer is a long way away perhaps an airline that is already operating the aircraft near to hand will have a suitable training facility, and this may be the alternative. If the aircraft are not bought new, but have been acquired from another airline, then it is quite possible that the vendor will carry out the necessary training programme as part of the transaction. The author has, in fact, taken part in such a programme where, in this instance, the

Fig. 13.1   Airport ground equipment in use. The airport is Salzburg, Austria. (Courtesy of Salzburger Flughafen GmbH.)

vendor airline delivered the aircraft to the customer by its training crews and then carried out the conversion and type rating instruction at the purchaser's own base.

Once the necessary crews have been fully trained to the standard required by the regulatory authorities, the aircraft may be put into service. The necessary checks will, or should, have been made of all equipment and facilities at every airport, including alternates, that will be visited regularly by the new aircraft. Nevertheless, it is still good practice to operate the aircraft on the intended route(s) for a flight without commercial pax. simply to check that it *does* fit in practice. And such flights may be included in the final training programme, perhaps in the form of line checks (which ensure that the captain knows how to make full use of both airborne and ground systems). It may be acceptable, too, that some 'invisible' return in non-tangible form may be forthcoming as a result of such flights – or proving flights. For example, passengers may be carried on a non-fare paying basis from the travel and aeronautical press, in the hope that free publicity, of a complimentary nature, will result. On normal routes, where little or nothing should go wrong, this is a useful practice. The kind of thing that must not happen is where the 'guests' witness a debacle, e.g. the airport's steps, or gate systems, do not fit the aircraft and the 'guests' cannot disembark. This results in quite the wrong sort of publicity, naturally (see Fig. 13.1).

Special Category Airports
Route proving flights are not required, by regulation, to simple, straightforward airports. Unfortunately, not all airports fall into this category, which can be referred to as category A airports. There are some airports that are not quite straightforward, and these may be referred to as being category B. Then there are a few 'difficult' airports, and these are referred to as being category C. In the case of cat. B airports, e.g. Gibraltar, additional training is required, together with the imposition of certain operational restrictions as regards the operating crew. For example, the flight crew should have special briefing regarding the destination, and a higher than normal AOM imposed, together with daylight-only operation, prior to their first visit. As in the case of cat. A airports, a proving flight is not required by regulation. However, it becomes somewhat more desirable, in the opinion of the author. Cat. C airports must be treated differently, and no captain may make an initial visit unless he is flying under supervision. But, one may well ask, what does the supervising pilot have that fits him, or her, for this task, and how was this asset acquired? It is all a little like the chicken and egg query, of course. But somebody has to have made a visit, in command, to the

Fig. 13.2   Arriving at a category B/C airport on ILS. No right turn is possible and the aircraft is just over the committed to land point. The missed approach calls for a turn to the left from this point but not after. The terrain ahead will not permit a straight ahead climb, with an engine out for many types. (Courtesy of Salzburger Flughafen GmbH.)

intended destination within the previous 13 months. Special briefing will also be a 'must' (see Fig. 13.2).

Unless there is a resident airline at the destination that can supervise the initial operation there is essentially one option. A proving and training flight must be made, for the express purpose of training, but including proving as well. Under the circumstances it is commercially sound sense for a sufficient number of pilots as is thought to be necessary (both captains and co-pilots) to take part in this flight, under the supervision of a training captain. Once the initial landing has been made, after an instrument approach, the training captain will have automatically checked himself as a qualified supervisory commander. He will also, it is likely, have carried out suitable simulator training himself before the initial flight. He can now carry out a programme of local training at the cat. C airport in question for the other commanders (and, if applicable, for the co-pilots), supervising IFR approaches and landings made by the other pilots, the captains at least having also received simulator training. What will be required must, of course, rest on the authorities controlling the operator. However, this flying will bring with it no tangible commercial benefits, such as publicity, although airline technical staff may participate to gain experience and to observe any likely problem areas. But the route may not be operated commercially unless and until the commanders have been route checked, and their training records annotated to this effect.

En Route

Irrespective of the destination airport's category, a new route can repay study from first-hand experience, with the objective of determining the optimum procedures to be followed. A typical example is finding the optimum route profile, from takeoff to landing, and the best power settings. If the destination airport is located in a mountainous region, and therefore likely to be cat. C, (e.g. Innsbruck) will the optimum profile be a climb to cruising FL based on the best distance:fuel:time criterion, followed by a cruise extended for as long as possible, and a steep descent? This is but an example and a question, not necessarily a recommended practice, of course. Again, as the destination authorities become better acquainted with the new operator they may well discuss, or even recommend, a variation in the arrival and return departure routings that could benefit the operator. So, while route proving can normally be expected to relate to the initial flight it may be advantageous to extend this, by specialist route observation afterwards.

14:   Sundry Considerations

In the preceding chapters an attempt has been made, in a condensed form, to outline the main functions and tasks of an airline's route planning organisation. It is not claimed that everything has been fully covered in detail, as the subject is wide and of considerable complexity. However, the objective has been to guide the reader into the main channels of this matter and to give an idea as to what may be encountered in practice. There are a few small points that, while of some importance, have not been dealt with or have only received a passing mention. An attempt will now be made to discuss these in the course of this concluding chapter.

Diplomatic Clearance

Normally, before either a scheduled service, or a charter flight is operated, it may well be necessary to obtain the approval of any foreign government involved in the case of an international operation. This is at least a courtesy, if in doubt, but may also be a requirement of that government on diplomatic grounds. If a foreign country requires a non-national airline to obtain prior approval to operate into, *or through* its national airspace, this will be made clear through such countries' Aeronautical Information Publication (AIP). Most Aeronautical Information Service (AIS) units in an airline's country of domicile carry copies of foreign AIPs and will advise as to other states' requirements in this context. Or the operator may hold a library of AIPs relevant to its sphere of operations. Some states have very strict requirements as regards the use of their airspace by foreigners, while others are quite satisfied with the despatch of a Flight Plan.

In those cases where diplomatic clearance is required, the procedure is laid down in the relevant AIP (RAC section). If a scheduled service is involved, the diplomatic niceties and requirements can be carried out at the same time as the fare structure negotiations mentioned earlier. A formal letter to the appropriate foreign authority, giving schedules, aircraft type, a photocopy of the licence (if required) can take care of the diplomatic clearance aspects. But, in the case of a charter flight, things usually need to be done more urgently, and the AIP will prescribe what action is necessary. Where diplomatic clearance is required it should be borne in mind that this *must* be obtained before the flight may be operated. The most normal requirement is that a prepaid reply cable is sent to an authority designated in the AIP, giving the name of the operator, aircraft type and registration, date(s) and times of operation, purpose of flight, and any relevant data or information. The cable signature will identify the applicant, with a cable address; normally ten words reply paid are sufficient. This cable should be sent as soon as the charter is firm, or on receipt of notification from the commercial department that it must be considered to be so. This early action is to allow for any questions that the foreign authority may wish to raise in an initial reply to be cleared up as soon as possible. Normally, unless the foreign government is characteristically difficult, a reply will be received similar to the following format: 'Your . . . of (date). Approved: Aeropolice (country of origin)'. It is just possible that the reply will be in the negative, usually because of some local situation, such as a state visit. But, assuming that approval is received, a photocopy of both the cable of request, and the reply thereto, should be filed by route planning and the originals included in the captain's flight brief. (The author has personal experience of an arrival at a foreign airport where, by a misunderstanding on the part of his planners, no diplomatic clearance had been applied for, although clearly indicated in the AIP as being a requirement. The experience was embarrassing! So, pay attention to the applicable diplomatic clearance requirements.

The Flight Brief
It is normally a requirement that complete operational data for regularly used sectors is included in the route book, or other elements of an airline's operations manual. But, on occasion, a single flight crops up that it not likely to recur. Clearly, if every such charter flight called for an entry of all data in the route book, this could in time make the volume unmanageable due to sheer bulk. So, for these 'one-off' flights it is normally acceptable for the operator to issue what is known as a flight brief. While being a separate, self-contained document, it is deemed to form an integral part of the airline's operations manual and as such it carries the same authority. A flight brief should contain the following data and information. It should contain the designated flight nos, schedule, and sector(s) to be operated. It should specify the alternate airport(s) to be used should the conditions at any destination airport be below the designated AOM. It should also give the charterer's name, address, telephone and telex numbers, and any other form of communications information. Information should also be provided regarding the contracted payload and in-flight services required. The handling agents at both the destination and alternate airport(s) should be specified, together

with their addresses and all communications facilities. Finally, under the general commercial data in the flight brief, nominated fuel suppliers will be noted for both destination(s) and alternate(s). Most probably, unless the charter is very sudden indeed, the nominated fuel suppliers will have been advised by their main office(s) as to the expected uplift requirements.

A flight brief *must* provide route data, airfield performance data, and AOM to the same standard as for routes contained in the operations manual, and such data must be shown to form part of this document. A convenient method of presenting such data is to use pro-formae to the same format as the relevant data would assume if it were to be presented in the main operations manual. Thus it is useful to keep a supply of printed 'blanks' for this purpose. What is required are plogs for each sector to be flown, including the return to base (or permanent out-station), a table of takeoff performance data for all destination and alternate airport's runways, e.g. if the route book provides 'D' values then the flight brief should do so likewise, and a table of AOM values for all aids that are in use at these airports, again in the same route book format. It is desirable that each sheet or plog that is issued should bear a note to the effect that it forms part of the route book volume of the operations manual. Each sheet should bear an attachment number that refers to the flight brief, and the number of attachments should be listed in the flight brief bearing each attachment number. Appendix 'A' shows a specimen flight brief in outline, or skeleton form, for reference.

Flight Time Limitations
All air carriers are required, by their various regulatory authorities, to limit the periods flight crews may be on duty *and* may fly within a given period. It is therefore of extreme importance that the charter arrangements are made bearing this in mind. The schedule that appears in the charter contract *must* comply with the requirements laid down by the airline's regulatory authority. Broadly speaking there is the duty period and the flying duty period. The first spans the period of time from when the flight crew reports before takeoff to a specified time after the landing. The charter schedule must be capable of meeting these requirements, and a realistic margin should be allowed. If the charter is a round trip, the whole must be subject to the appropriate time limitations, although if the outbound and return sectors are short enough, and if suitable rest facilities for the crew are available during the interval at the destination, some easing of the time limits may be acceptable. What is *not* permissible is for a schedule to be issued that either exceeds the total 'duty day' or avoids this only marginally.

It may be that the charter is to a destination with the scheduled return to be made a day or so after. In such cases, so long as the flight time limits are met on both the outbound and return flights, the crew will night stop and thus meet the rest requirements. It must therefore be notified to the commercial department, when supplying the block time and fuel information, if the proposed charter pattern meets the flight time limitations or not. If not, an acceptable schedule must be agreed that does, and this could well influence the charter costs, and therefore price.

If the proposed charter involves a long stage, using a large aircraft, it may be that extra crew may be carried. If the crew is augmented for the purpose of extending flight time limitations the aircraft must be fitted with acceptable rest accommodation, in the form of bunks, not seats. This may affect the APS weight and therefore the payload available. The turn-round time at the destination must also allow the whole extended time crew a statutory rest period before any return or onward flight is made. Or a 'slip crew' must be available, rested, to fly the return flight. (Incidentally, it must be made clear that the requirements relating to flight time limitations apply equally to scheduled services as well as charter flights.) One final point as regards crew complement. Certain designated areas across the world require the carriage of a specialist flight navigator, before any part of them may be overflown. But, if the aircraft is equipped with acceptable navigational aids, i.e. inertial, or area navigation systems based on radio transmissions, such as Omega, most, if not all authorities, will accept such installations in lieu of the navigator.

We have been discussing the requirements as regards the flight crew, but the cabin crew are also subject to requirements aimed at the prevention of fatigue, and also the minimum complement of a cabin crew. This is normally based on the number of passenger seats fitted, and the various authorities require an acceptable ratio of cabin staff to passengers. However, this will normally be allowed for in the APS weight and it is not normally the case for extra cabin staff to be carried other than to provide a higher than normal standard of cabin service, as may be dictated by the charter agreement.

Appendix A: Specimen Outline Flight Brief

ROCKALL INTERNATIONAL AIRLINES

FLIGHT BRIEF

Flight Brief No. 176/89   Flight No. RI 2702/3
              From: Belfast (Aldergrove) (BFS) to Larnaca (LCA)
Date Out: April 20 1989   Date of Return Dep: April 21 1989
Aircraft: Starshooter – 200
Schedule: 1000 ↓ Belfast (EGAA)   ↑ 1445
                  1520 ↓ Larnaca (LCLK)   ↑ 0900
                  (All times are GMT)
Alternate: Paphos International (LCPH) Notice given for use as
                  Alternate and Approved.
Payload: 125 Pax. each way.
Catering: 123 'Y+', 2 vegetarian. Free bar. Duty free goods on sale basis.
Charterer: Arrow Tourist Services,
                  23 Tite Walk,
                  Rochester, Kent. Tel: Rochester 23456. Telex: 65432 ATS G
Handling Agent: Speedy Handling Services,
                  Larnaca Airport, Cyprus, Tel: Larnaca 7272900 Telex: 2727
                  SPDY C
                  Paphos International Airport, Cyprus, Tel: Paphos 22221
                  Telex: 2929 SPDY C
Diplomatic Clearance: Granted. Original Clearance attached (Attachment 5).
Fuel supplier: Destination and Alternate – Shell.
Operations data: Plogs, Out and Return – Attachments 1 and 2.
                  Airport Performance – No WAT Limits, 'D' values and $V_1:V_R$
                                       as tabulated in Attachment 3.
                  AOM – As detailed in Attachment 4.
This Flight Brief, and Attachments 1 to 5 incl. form part of the Operations
Manual.

Note
For illustration purposes only. Names of Charterer and Handling Agent fictitious. Fuel
supplier may, or may not actually be at these airports.

Appendix B: UK Air Navigation Order and ICAO Definitions

UK Air Navigation Order Definitions

These definitions refer to UK AOM methodology, a description of which follows:

Aerodrome   Any area of land or water designed, equipped, set apart or commonly used for affording facilities for the landing and departure of aircraft and includes any area of space, whether on the ground, on the roof of a building or elsewhere, which is designed, equipped or set apart for affording facilities for the landing and departure of aircraft capable of descending or climbing vertically, but shall not include any area the use of which for affording facilities for the landing and departure of aircraft has been abandoned and has not been resumed.

Aerodrome operating minima (AOM)   In relation to the operation of an aircraft at an aerodrome means the cloud ceiling and runway visual range for takeoff, and the decision height or minimum descent height, runway visual range and visual reference for landing, which are the minimum for the operation of that aircraft at that aerodrome.

Aeronautical radio station   A radio station on the surface, which transmits or receives signals for the purpose of assisting aircraft.

Approach to landing   That portion of the flight of the aircraft, when approaching to land, in which it is descending below a height of 1000 ft above the relevant specified decision height or minimum descent height.

Certificate of airworthiness   Includes any validation thereof and any flight manual, performance schedule or other document, whatever its title, incorporated by reference in that certificate relating to the certificate of airworthiness.

Cloud ceiling (CC)   In relation to an aerodrome means the vertical distance from the elevation of the aerodrome to the lowest part of any cloud visible from the aerodrome which is sufficient to obscure more than one-half of the sky so visible.

Commander   In relation to an aircraft means the member of the flight crew designated as commander of that aircraft by the operator thereof, or, failing such a person, the person who is for the time being the pilot in command of the aircraft.

Competent authority   In relation to the United Kingdom, the Authority, and in relation to any other country, the authority responsible under the law of that country for promoting the safety of civil aviation.

Contracting State   Any State (including the United Kingdom) which is party to the Convention on International Civil Aviation signed on behalf of the Government of the United Kingdom at Chicago on the 7 December 1944.

Controlled airspace   Control areas and control zones.

Control area   Airspace which has been notified as such and which extends upwards from a notified altitude or flight level.

Control zone   Airspace which has been notified as such and which extends upwards from the surface.

Decision height (DH)   In relation to the operation of an aircraft at an aerodrome means the height in a precision approach at which a missed approach must be initiated if the required visual reference to continue that approach has not been established.

Flight level (FL)   One of a series of levels of equal atmospheric pressure, separated by notified intervals and each expressed as the number of hundreds of feet which would be indicated at that level on a pressure altimeter calibrated in accordance with the International Standard Atmosphere and set to 1013.2 millibars.

Flight visibility   The visibility forward from the flight deck of an aircraft in flight.

Instrument flight rules (IFR)   Instrument flight rules prescribed under Article 64(1) of this Order.

Instrument meteorological conditions (IMC)   Weather precluding flight in compliance with the visual flight rules.

Maximum total weight authorised (MTWA)   In relation to an aircraft means the maximum total weight of the aircraft and its contents at which the aircraft may take off anywhere in the world, in the most favourable circumstances in accordance with the certificate of air-worthiness in force in respect of the aircraft.

Minimum descent height (MDH)   In relation to the operation of an aircraft at an aerodrome means the height in a non-precision approach below which descent may not be made without the required visual reference.

Nautical mile   The International Nautical Mile, that is to say, a distance of 1852 metres.

Non-precision approach   An instrument approach using non-visual aids for guidance in azimuth or elevation but which is not a precision approach.

Notified   Set forth in a document published by the Authority and entitled 'United Kingdom Notam' or 'United Kingdom Air Pilot' and for the time being in force.

Pilot in command   In relation to an aircraft means a person who for the time being is in charge of the piloting of the aircraft without being under the direction of any other pilot in the aircraft.

Precision approach   An instrument approach using instrument landing system, microwave landing system or precision approach radar for guidance in both azimuth and elevation.

Runway visual range (RVR)   In relation to a runway means the distance in the direction of takeoff or landing over which the runway lights or surface markings may be seen from the touchdown zone as calculated by either human observation or instruments in the vicinity of the touchdown zone or where this is not reasonably practicable in the vicinity of the midpoint of the runway; and the distance, if any communicated to the commander of an aircraft by or on behalf of the person in charge of the aerodrome as being the runway visual range shall be taken to be the runway visual range for the time being.

ICAO Definitions

Aerodrome elevation   The elevation of the highest point of the landing area [see *Elevation*].

Altitude   The vertical distance of a level, a point, or an object considered as a point, measured from mean sea level (MLS).

Arrival routes   Routes identified in an instrument approach procedure by which aircraft may proceed from the en route phase of flight to an initial approach fix.

Base turn   A turn executed by the aircraft during the initial approach between the end of the outbound track and the beginning of the intermediate or final approach track. The tracks are not reciprocal.
Note   Base turns may be designated as being made either in level flight or while descending, according to the circumstances of each individual procedure.

Circling approach   An extention of an instrument approach procedure which provides for visual circling of the aerodrome prior to landing.

Decision altitude/height(DA/H)   A specified altitude or height (A/H) in the precision approach at which a missed approach must be initiated if the required visual reference to continue the approach has not been established.
Note 1   Decision altitude (DA) is referenced to mean sea level (MSL) and decision height (DH) is referenced to the threshold elevation.
Note 2   The required visual reference means that section of the visual aids or of the approach area which should have been in view for

sufficient time for the pilot to have made an assessment of the aircraft position and rate of change of position, in relation to the desired flight path.

DME distance   The line of sight distance (slant range) from the source of a DME signal to the receiving antenna.

Elevation   The vertical position of a point or a level, on or affixed to the surface of the earth, measured from mean sea level.

Final approach segment   That segment of an instrument approach procedure in which alignment and descent for landing are accomplished.

Flight level (FL)   A surface of constant atmospheric pressure which is related to a specific pressure datum, 1013.2 hPa [millibars or 29.92 in. Hg] and is separated from other such surfaces by specific pressure intervals.

Note 1   A pressure type altimeter calibrated in accordance with the Standard Atmosphere:

(a) when set to a QNH altimeter setting, will indicate altitude;

(b) when set to a QFE altimeter setting, will indicate height above the QFE reference datum; and

(c) when set to a pressure of 1013.2 hPa, [millibars or 29.92 in. Hg] may be used to indicate flight levels.

Note 2   The terms height and altitude, used in Note 1 above, indicate altimetric rather than geometric heights and altitudes.

Heading   The direction in which the longitudinal axis of an aircraft is pointed, usually expressed in degrees from North (true, magnetic, compass or grid).

Height   The vertical distance of a level, a point or an object considered as a point, measured from a specified datum.

Holding procedure   A predetermined manoeuvre which keeps an aircraft within a specified airspace while awaiting further clearance.

Initial approach segment   That segment of an instrument approach procedure between the initial approach fix and the intermediate approach fix or, where applicable, the final approach fix or point.

Instrument approach procedure [IAP]   A series of predetermined manoeuvres by reference to flight instruments with specified protection from obstacles from the initial approach fix, or where applicable, from the beginning of a defined arrival route to a point from which a landing can be completed and thereafter, if a landing is not completed, to a position at which holding or en route obstacle clearance criteria apply.

Intermediate approach segment   That segment of an instrument approach procedure between either the intermediate approach fix and the final approach fix or point, or between the end of a reversal, race track or dead reckoning track procedure and the final approach fix or point, as appropriate.

Level   A generic term relating to the vertical position of an aircraft in flight and meaning variously, height, altitude or flight level.

Minimum descent altitude/height (MDA/H)   A specified altitude/height in a non-precision approach or circling approach below which descent may not be made without visual reference.

Minimum sector altitude   The lowest altitude which may be used under emergency conditions which will provide a minimum clearance of 1000 ft above all objects located in an area contained within a sector of a circle of 25 nm radius centred on a radio aid to navigation.

Missed approach point (MAPt)   That point in an instrument approach procedure at or before which the prescribed missed approach procedure must be initiated in order to ensure that the minimum obstacle clearance is not infringed.

Missed approach procedure   The procedure to be followed if the approach cannot be continued.

Obstacle assessment surface (OAS)   A defined surface intended for the purpose of determining those obstacles to be considered in the calculation of obstacle clearance altitude/height for a specific ILS facility and procedure.

Obstacle clearance altitude/height (OCA/H)   The lowest altitude (OCA), or alternatively the lowest height above the elevation of the relevant runway threshold or above the aerodrome elevation as applicable (OCH), used in establishing compliance with appropriate obstacle clearance criteria.

Note   See [I.C.A.O. PANS-OPS 8168/611, Vol. I] Part III, Chapter 1, 1.5.1 for specific applications of this definition.

Precision approach procedure   An instrument approach procedure utilizing azimuth and glide path information provided by ILS or PAR.

Primary area   A defined area symmetrically disposed about the nominal flight track in which full obstacle clearance is provided (see also *Secondary area*).

Procedure turn   A manoeuvre in which a turn is made away from a designated track followed by a turn in the opposite direction to permit the aircraft to intercept and proceed along the reciprocal of the designated track.

Note 1   Procedure turns are designated 'left' or 'right' according to the direction of the initial turn.

Note 2   Procedure turns may be designated as being made either in level flight or while descending, according to the circumstances of each individual instrument approach procedure.

Racetrack procedure   A procedure designed to enable the aircraft to reduce altitude during the initial approach segment and/or establish the aircraft inbound when the entry into a reversal procedure is not practical.

Reversal procedure   A procedure designed to enable aircraft to reverse direction during the initial approach segment of an instrument approach procedure. The sequence may include procedure turns or base turns.

Secondary area   A defined area on each side of the primary area located along the nominal flight track in which decreasing obstacle clearance is provided (see also *Primary area*).

Threshold (THR)   The beginning of that portion of the runway usable for landing.

Track   The projection on the earth's surface of the path of an aircraft, the direction of which path at any point is usually expressed in degrees from North (true, magnetic or grid).

Transition altitude   The altitude at or below which the vertical position of an aircraft is controlled by reference to altitudes.

Transition layer   The airspace between the transition altitude and the transition level.

Transition level   The lowest flight level available for use above the transition altitude.

Visual manoeuvring (circling area)   The area in which obstacle clearance should be taken into consideration for aircraft carrying out a circling approach.

Glossary

Aerad   A flight guide for navigational purposes published by British Airways.

Airline   An air transport undertaking or operator.

Airways   Designated 'corridors' for use by aircraft in transit and marked at intervals by radio navigational aids, on or off the airway. Normally 10 nm wide, but with varying upper and lower limits, which are promulgated. Airways are controlled airspace (in the UK, and certain states that follow UK practices, many airlines misuse the word 'airways').

Aeronautical Information Publication (AIP)   A state-published official document that contains navigational and aerodrome information, and regulatory data, pertaining to that state's airspace. Still referred to, in the UK, as the 'Air Pilot'.

Aeronautical Information Service (AIS)   An air navigation and aerodrome information service normally provided by a state.

Air Traffic Control (ATC)   A ground-based organisation covering each state's airspace that controls the safe passage of aircraft traffic by allotting altitudes and separation between every aircraft within that state's airspace.

Alternate   An airport designated before the departure of a flight as being an alternative point of landing should conditions at the intended destination of unsuitable for a landing to be made.

Altitude   The vertical distance of a level, an object, or a point measured from mean sea level (msl).

APS Weight   The weight of an aircraft prepared for service, normally less only the fuel and payload to be loaded.

APU   Auxiliary Power Unit.

AOM   Aerodrome Operating Minima. Declared or approved minimum values of cloud base and visibility at an aerodrome for takeoff and landing.

ASD/ASDA   Accelerate Stop Distance/Available. The distance required, or available, for an aircraft taking off to accelerate to a decision point associated with the recognition of an engine failure and then to abandon (abort) the takeoff and come to a halt with no more than superficial damage.

Burnoff   The amount of fuel burned, or expected to be burned, on any given flight. May be calculated or measured from starting engines to shutting down, or from start of takeoff to end of landing.

Base   The location of an airline's main operating centre (i.e. aerodrome).

Bug   A small moveable index located on an aircraft's instrument(s) to identify significant values.

CAA   Civil Aviation Authority (UK).

Circling height   The minimum height at which an aerodrome may be circled visually prior to landing.

Circling minima   The minimum values of cloud base and visibility at which an aerodrome may be circled visually.

Checks   Specified intervals of time, either in flight hours or calendar time, after which scheduled maintenance work must be carried out.

Clearway   A specified area at the end of the takeoff runway that is free of any obstacles that would penetrate a 1.25% horizontal plane.

'D' value   A method of indicating the equivalent balanced field length available or required.

Decision – altitude/height   A specified height or altitude for a precision approach at which a missed approach must be initiated if the required visual reference to continue the approach has not been attained.

Decision point   A point during the takeoff at which, following recognition of an engine failure, the takeoff may be either abandoned or continued within the confines of the declared runway values available.

Elevation   The vertical distance of a point or object on the earth measured vertically from mean sea level.

ED/EMD   The required distance, or distance promulgated as being available, for carrying out an abandoned take off (see *Decision point*).

En Route Alternate (ERA)   An alternate aerodrome en route to the intended destination and which may be designated as an alternate, subject to certain requirements being observed.

FAA   Federal Aviation Administration (USA).

Flight brief   A document provided for flight crew use containing essential data not already contained in the operator's operations manual. Normally issued to cover a single charter flight, or flights made at short notice. It will also usually contain details of the schedule to be followed and administrative information. The flight brief is normally deemed to form part of the operations manual.

Flight Level (FL)   An altitude or altitudes measured by an aircraft's altimeter having its subscale set to 1013.2 millibars or 29.92 in. Hg and having values at exact thousands of feet. FLs are expressed by 2 or 3 digits, e.g. FL280 = 28 000 ft in altimetric terms rather than geographical. These values are based on constant atmospheric pressure appropriate to the values given above.

Flight manual (AFM)   A document forming part of an aircraft's certificate of airworthiness, containing scheduled and approved performance data and procedures.

Flight path   The profile of an aircraft's takeoff, up to the point at which it assumes the en route configuration, e.g. wheels and flaps retracted, climb power. An engine is assumed to have failed at the decision point (see above).

Flight plan   A document completed prior to takeoff and filed with ATC giving details of the proposed route, altitudes/Fls, alternate aerodromes, endurance, number of persons on board, emergency equipment, etc.

Flight planning   The preparation of data for the compilation of the flight plan.

Fuel price index   The cost of fuel at aerodromes away from base expressed as a percentage of base cost.

Flying Staff Instructions (FSI)   A formal document giving flight crews essential data or procedures to be followed, together with an operator's policies. May be permanent (in which case it will be incorporated into the appropriate part of the operations manual) or temporary.

Height   The vertical distance of a level, point, or object from a specified datum, e.g. aerodrome elevation.

Holding   A designated flight procedure, located by a radio navigation aid, that is used in cases when a direct approach and landing cannot be immediately initiated. Normally flown at long range cruise speed or a designated value of speed appropriate to the aircraft type.

International Civil Aviation Organisation (ICAO)   A specialised United Nations agency that issues *advisory* material on a global basis.

Instrument Landing System (ILS)   A precision landing aid using radio signals from a ground installation.

Jet Pipe Temperature (JPT)   The temperature of the gases in a jet or turbine engine, measured at the exhaust. Also known as EGT, or exhaust gas temperature.

Jeppesen   An alternative flight guide to Aerad.

Landing Weight (LW)   The weight of an aircraft at touchdown. Required LW (Reqd.LW) is the landing weight that does not produce any weight loss. Regulated LW is the maximum landing weight permitted by the landing distance available (LDA), while the max. LW is the maximum weight authorised in the AFM for landing (MLWA).

Limitations   Certain values or parameters specified in the AFM which may not be exceeded e.g. minimum start-up temperature.

Location indicator   An ICAO group of four letters that identifies an aerodrome or facility, e.g. EGLL = London (Heathrow), EGTT =

London ATC Centre.

MLWA   See landing weight.

M_{MO} Maximum operating Mach no.

Minimum Safe Altitude (MSA)   A specified altitude that provides a minimum of 1000 ft vertical clearance and 25 nm lateral clearance from obstacles within a declared area or sector.

Minimum Sector Fuel (MSF)   The minimum amount of fuel that may be specified by an operator to be loaded for any particular flight, irrespective of the forecast en route winds. MSF may be exceeded, but not reduced.

Maximum Takeoff Weight Authorised (MTWA)   The maximum weight permitted by the AFM for an aircraft at the start of the takeoff run (see *Takeoff weight*).

Maximum Zero Fuel Weight (MZFW)   A structural weight, given in the AFM, that may only be exceeded by the weight of fuel, up to MTWA.

N_1 *and* N_2 The rotational speed of the low and high pressure compressor of a turbine engine. Expressed normally as a percentage of the maximum permitted value, but may be expressed as rpm.

Net flight path   The takeoff flight profile as presented in the AFM and including factors as safeguards (see *Flight path*).

Non-precision approach   Any approach using a radio aid other than ILS (with glide path) or PAR.

Outside air temperature (OAT)   The free air, or static, ambient temperature.

Obstacle Clearance Limit (OCL)   Now being phased out, an OCL is a vertical value above airfield level and based on the highest obstacle within the defined area for an instrument approach using a particular aid. It is normally promulgated by the state authority.

Obstacle Clearance Altitude (OCA) or Height (OCH)   The lowest altitude (OCA), or height (OCH) above the appropriate runway threshold elevation or above aerodrome elevation (OCH) used to obtain the required obstacle clearance.

Operations manual   The flight crew's bible. A document published by an operator and containing the operator's policies, responsibilities, regulatory information, flight planning data etc.

PAR   Precision approach radar.

Plog   An abbreviation for prepared navigation and fuel log, on which full route details are included, with standard times for specified values of wind component and temperature. Rapidly becoming a computerised process.

P_1 *and* P_7 An alternative engine power measurement to N_1/N_2, using pressures instead of rpm. See EPR and N_1 and N_2 above.

Precision approach   An instrument approach made using full ILS, or PAR only.

Ratio V_1:V_r   Controlled by the ASDA this is the relationship between the rotation speed (which is controlled by weight) and the decision speed (which is controlled by the ASDA also, but associated with the TODA). See *Decision point* and *ED/EMD*.

Reserve fuel   The fuel required to divert following a missed approach at the intended destination to a designated alternate and to then hold for a specified time and then carry out an approach and landing.

Reporting Point (RP)   A geographical point, normally on an Airway (see above) and located by means of radio navigational means at which ATC will require a report from overhead (unless instructed otherwise).

RLW/REQ. LW   See *Landing weight*.

Rejected takeoff (RTO)   An abandoned, or aborted takeoff. See *ASD*.

Route book   A volume of the operations manual (see above) containing navigational and flight planning data, AOM, performance, emergencies, and so on.

Route licence   A licence granted by state authorities permitting the operation of an air transport service, but subject to the satisfaction of the appropriate authorities as regards safety standards. A route licence, if international, may require the approval of the external state(s) involved as well.

RTOW/REQD. TOW   See *Takeoff weight*.

Runway   Strictly speaking, the paved or prepared takeoff surface available for an aircraft to become airborne, allowing for engine failure. For performance purposes a runway may comprise this element, plus an extension engineered to a lower standard specifically available for deceleration only following an aborted takeoff and known as stopway, plus a declared distance having a 1.25% horizontal plane free of obstacles for accommodating the after-lift off climb and known as clearway. A runway may comprise the prepared run alone (TOR), or it may comprise TOR + Stopway = ASDA, or TOR + Clearway = TODA or it may contain all three distances, thus providing TORA, ASDA, and TODA. See *ASDA*. All these values are normally promulgated by the appropriate authorities.

Runway visual range   An element of AOM (see above) covering the horizontal visibility along the runway.

Standard Instrument Departure (SID)   A prescribed departure route from an aerodrome as issued by the relevant authorities.

Standard Terminal Arrival Route (STAR)   Similar to a SID but relating to arrivals.

Standard weights   The notional average weights of males, females, and children used for estimating the loaded weight of an aircraft.

Takeoff Weight (TOW)   The weight of an aircraft at the commencement of the takeoff run. The regulated TOW (RTOW) is the maximum value permitted by runway, or altitude and temperature reasons, whichever is the lower. The reqd. TOW is the weight of an aircraft loaded with payload and the sector fuel. It may not exceed the RTOW.

Target Threshold Speed (TTS) V_{AT_1} (1 engine out); V_{AT_0} (all engines).

Transponder   An airborne aid to identification, and providing certain flight data. Used in association with secondary ground radar, by which it is 'interrogated' and automatically replies.

Turbine Gas Temperature (TGT)   See *Jet Pipe Temperature*.

Type Rating Examiner (TRE)   A pilot authorised by the authorities to carry out periodical examinations and tests of other pilots concerning their competency to operate a specified aircraft type. A TRE will also conduct initial examinations and tests to qualify other pilots to operate such an aircraft and to include the type rating in their licences.

Weight-Altitude-Temperature (WAT)   A limitation that can override the weights for takeoff and landing permitted by a runway. It is based on the climb performance of an aircraft with flaps and undercarriage extended as appropriate to takeoff or a missed approach, normally with one engine inoperative (for missed approach), but always so for takeoff.

Waypoint   A designated point on an airway or route that is normally marked by ground radio navigational means. A waypoint may or may not be a RP (see above).

Wind Component (W/C)   The value of the wind force in knots relative to the angle of the track being flown, e.g. a wind from 90° to the track produces a W/C = 0, while a similar wind blowing from a direction 180° to the track produces a W/C equal to the speed of the wind.

Weight statement   The weight of a fully equipped aircraft without crew, catering, passenger service items, payload, or usable fuel.

Zero Fuel Weight (ZFW)   The weight of an aircraft exclusive *only* of fuel load. See *Max. Zero Fuel Weight*.

Abbreviations

The following abbreviations are arranged by subject.

Altitude

DA	Density Altitude. The theoretical density of a standard atmosphere at that altitude.
PA	Pressure Altitude. The altitude shown on an altimeter when the sub-scale is set to 1013.2 mb or 29.92 in. Hg.
TA	True Altitude. The absolute altitude, or as close to this as is possible, using instrumentation fitted to an aircraft.

Speed

CAS	Calibrated Air Speed. This is equal to ASIR, corrected for both position and instrument errors. (CAS = TAS at msl at ISA.)
EAS	Equivalent Air Speed. This is equal to ASIR, corrected for position error, instrument error, and compressibility for altitude. (EAS = CAS at msl at ISA.)
IAS	Indicated Air Speed corrected for instrument error only.
TAS	True Air Speed. Speed relative to the outside air.
V_{EF}	The speed at which an engine fails, or is assumed to fail.
V_1	Takeoff Decision Speed.
V_R	Rotation Speed. The speed when the aircraft taking off assumes a position for flight (lift-off).
V_2	Takeoff safety speed.
V_3	Initial steady climb speed, all-engines operating at screen height for V_2 (35 ft).
V_4	Steady climb speed, with undercarriage up and flaps at takeoff.
V_{FR}	Flap retraction safety speed.
V_{FT_0}	Final takeoff speed.
V_{AT_1}	Target threshold speed – one engine out.

V_{AT_0}	Target threshold speed – all engines.
V_S	Stalling speed for the appropriate configuration.
V_{S_0}	Stalling speed for approach flaps, undercarriage up.
V_{S_1}	Stalling speed for any case under consideration.

Weight

MTOW	Maximum Permitted Takeoff Weight, structural.
MTWA	Maximum Takeoff Weight Authorised.
MLW	Maximum Landing Weight, structural.
MLWA	Maximum Landing Weight Authorised.
MZFW	Maximum weight of aircraft excluding fuel. This is a limiting weight and any increase may only comprise fuel.
RTOW	Regulated Takeoff Weight. The maximum takeoff weight permitted by runway, altitude, temperature, or obstacle clearance.
RLW	The maximum landing weight permitted by runway and altitude, or by approach climb WAT requirements.
Max. auw	Maximum all-up weight (similar to MTOW).
OWE	Operating Weight Empty.

Temperature

OAT	Outside Air Temperature. The temperature outside the aircraft in undisturbed air.
T_{FLEX}	The temperature used when calculating power reduction for Reduced Thrust takeoff.

Meteorological

AOM	Aerodrome Operating Minima.
CC	Cloud Ceiling.
DA	Descent Altitude.
DH	Descent Height.
MDA	Minimum Descent Altitude.
MDH	Minimum Descent Height.
FAF	Final Approach Fix.
IFV	In Flight Visibility.
MOC	Minimum Obstacle Clearance.
OCA	Obstacle Clearance Altitude.
OCH	Obstacle Clearance Height.

OCL	Obstacle Clearance Limit.
RVR	Runway Visual Range.
W/C	Wind Component.
W/V	Wind Velocity.

Runway Distances

TORA	Takeoff Run available.
EMDA/ASDA	Accelerate-Stop Distance Available.
TODA	Takeoff Distance Available.
LDA	Landing Distance Available.
TORR	Takeoff Run Required.
EMDR/ASDR	Accelerate-Stop Distance Required.
TODR	Takeoff Distance Required.
LDR	Landing Distance Required.

Engine Parameters

EPR	Engine Pressure Ratio
Pt_2	Inlet Pressure.
Pt_7	Exhaust Pressure.
N_1	Low pressure Compressor rpm.
N_2	High pressure Compressor rpm.
rpm	revolutions per minute.
TBO	Time Between Overhauls.
EGT	Exhaust Gas Temperature.
TGT	Turbine Gas Temperature.

General

JAR 25	European Joint Airworthiness Requirements (Part 25).
FAR 25	Federal Aviation Regulations (Part 25).
BCARs	British Civil Airworthiness Requirements.
CRZ	Cruise.
amsl	above mean sea level.
S.G.	Specific Gravity.
WED	Water Equivalent Depth.
ISWL	Isolated Single Wheel Load.
LCN	Load Classification Number.
psi	pounds per square inch.
NPRM	Notice of Proposed Rule Making (FAA).
C of G	Centre of Gravity.

nm	Nautical mile.
ILS	Instrument Landing System. A precision radio approach and landing aid which, where accurate enough, may be cleared for allowing a suitably equipped aircraft to land automatically, using its autopilot.
PAR	Precision Approach Radar (once known as GCA or Ground Controlled Approach). A precision approach and landing aid, mainly used by military airfields.
MLS	Microwave Landing System. A development of ILS, but not yet in any significant use.
VOR	Very High Frequency Omnidirectional Radio Range.
SRA	Search Radar. A non-precision radar approach aid.
VDF	VHF Direction Finder. A non-precision approach aid.
NDB	Non Directional Beacon. A non-precision radio approach aid, or cloud break aid, depending upon its location relative to the landing runway.
LLZ	Localiser. The azimuth guidance element of ILS. Non-precision without G/P.
G/P	Glide Path. The vertical guidance element of ILS.
FL	Flight Level.
ETA	Expected Time of Arrival.
nampkg	nautical air miles per kg.
fpm	feet per minute.
G/S	Ground Speed.
IFR	Instrument Flight Rules.
CVR	Cockpit Voice Recorder.
FMS	Flight Management Systems.
FDR	Flight Data Recorder.
u/c	undercarriage.
P/L	Payload. The load, excluding fuel, that an aircraft may carry for commercial or other purposes. The max. P/L that can ever be carried is the MZFW–APS weight.

Index